Students, families, and educators:

Thank you for being part of the *Eureka Math®* community, where we celebrate the joy, wonder, and thrill of mathematics.

In *Eureka Math* classrooms, learning is activated through rich experiences and dialogue. That new knowledge is best retained when it is reinforced with intentional practice. The *Learn, Practice, Succeed* book puts in students' hands the problem sets and fluency exercises they need to express and consolidate their classroom learning and master grade-level mathematics. Once students learn and practice, they know they can succeed.

What is in the Learn, Practice, Succeed *book?*

Fluency Practice: Our printed fluency activities utilize the format we call a Sprint. Instead of rote recall, Sprints use patterns across a sequence of problems to engage students in reasoning and to reinforce number sense while building speed and accuracy. Sprints are inherently differentiated, with problems building from simple to complex. The tempo of the Sprint provides a low-stakes adrenaline boost that increases memory and automaticity.

Classwork: A carefully sequenced set of examples, exercises, and reflection questions support students' in-class experiences and dialogue. Having classwork preprinted makes efficient use of class time and provides a written record that students can refer to later.

Exit Tickets: Students show teachers what they know through their work on the daily Exit Ticket. This check for understanding provides teachers with valuable real-time evidence of the efficacy of that day's instruction, giving critical insight into where to focus next.

Homework Helpers and Problem Sets: The daily Problem Set gives students additional and varied practice and can be used as differentiated practice or homework. A set of worked examples, Homework Helpers, support students' work on the Problem Set by illustrating the modeling and reasoning the curriculum uses to build understanding of the concepts the lesson addresses.

Homework Helpers and Problem Sets from prior grades or modules can be leveraged to build foundational skills. When coupled with *Affirm®*, *Eureka Math*'s digital assessment system, these Problem Sets enable educators to give targeted practice and to assess student progress. Alignment with the mathematical models and language used across *Eureka Math* ensures that students notice the connections and relevance to their daily instruction, whether they are working on foundational skills or getting extra practice on the current topic.

Where can I learn more about Eureka Math *resources?*

The Great Minds® team is committed to supporting students, families, and educators with an ever-growing library of resources, available at eureka-math.org. The website also offers inspiring stories of success in the *Eureka Math* community. Share your insights and accomplishments with fellow users by becoming a *Eureka Math* Champion.

Best wishes for a year filled with "aha" moments!

Jill Diniz

Jill Diniz
Chief Academic Officer, Mathematics
Great Minds

Eureka Math®
Grade 7
Module 3

Published by Great Minds®

Copyright © 2019 Great Minds®.

Printed in the U.S.A.

This book may be purchased from the publisher at eureka-math.org.
4   5   6   7   8   9   10   LSC   26   25   24   23   22   21

ISBN 978-1-64054-974-6

G7-M3-LPS-05.2019

Contents

Module 3: Expressions and Equations

Opening Exercise

Each envelope contains a number of triangles and a number of quadrilaterals. For this exercise, let t represent the number of triangles, and let q represent the number of quadrilaterals.

a. Write an expression using t and q that represents the total number of sides in your envelope. Explain what the terms in your expression represent.

b. You and your partner have the same number of triangles and quadrilaterals in your envelopes. Write an expression that represents the total number of sides that you and your partner have. If possible, write more than one expression to represent this total.

c. Each envelope in the class contains the same number of triangles and quadrilaterals. Write an expression that represents the total number of sides in the room.

d. Use the given values of t and q and your expression from part (a) to determine the number of sides that should be found in your envelope.

e. Use the same values for t and q and your expression from part (b) to determine the number of sides that should be contained in your envelope and your partner's envelope combined.

f. Use the same values for t and q and your expression from part (c) to determine the number of sides that should be contained in all of the envelopes combined.

g. What do you notice about the various expressions in parts (e) and (f)?

Example 1: Any Order, Any Grouping Property with Addition

a. Rewrite $5x + 3x$ and $5x - 3x$ by combining like terms.

Write the original expressions and expand each term using addition. What are the new expressions equivalent to?

b. Find the sum of $2x + 1$ and $5x$.

Lesson 1:   Generating Equivalent Expressions

c. Find the sum of $-3a + 2$ and $5a - 3$.

Example 2: Any Order, Any Grouping with Multiplication

Find the product of $2x$ and 3.

Example 3: Any Order, Any Grouping in Expressions with Addition and Multiplication

Use any order, any grouping to write equivalent expressions.

a. $3(2x)$

b. $4y(5)$

c. $4 \cdot 2 \cdot z$

d. $3(2x) + 4y(5)$

e. $3(2x) + 4y(5) + 4 \cdot 2 \cdot z$

f.   Alexander says that $3x + 4y$ is equivalent to $(3)(4) + xy$ because of any order, any grouping. Is he correct? Why or why not?

Relevant Vocabulary

VARIABLE (DESCRIPTION): A *variable* is a symbol (such as a letter) that represents a number (i.e., it is a placeholder for a number).

NUMERICAL EXPRESSION (DESCRIPTION): A *numerical expression* is a number, or it is any combination of sums, differences, products, or divisions of numbers that evaluates to a number.

VALUE OF A NUMERICAL EXPRESSION: The *value of a numerical expression* is the number found by evaluating the expression.

EXPRESSION (DESCRIPTION): An *expression* is a numerical expression, or it is the result of replacing some (or all) of the numbers in a numerical expression with variables.

EQUIVALENT EXPRESSIONS: Two expressions are *equivalent* if both expressions evaluate to the same number for every substitution of numbers into all the letters in both expressions.

AN EXPRESSION IN EXPANDED FORM: An expression that is written as sums (and/or differences) of products whose factors are numbers, variables, or variables raised to whole number powers is said to be in *expanded form*. A single number, variable, or a single product of numbers and/or variables is also considered to be in expanded form. Examples of expressions in expanded form include: 324, $3x$, $5x + 3 - 40$, and $x + 2x + 3x$.

TERM (DESCRIPTION): Each summand of an expression in expanded form is called a *term*. For example, the expression $2x + 3x + 5$ consists of three terms: $2x$, $3x$, and 5.

COEFFICIENT OF THE TERM (DESCRIPTION): The number found by multiplying just the numbers in a term together is the *coefficient of the term*. For example, given the product $2 \cdot x \cdot 4$, its equivalent term is $8x$. The number 8 is called the coefficient of the term $8x$.

AN EXPRESSION IN STANDARD FORM: An expression in expanded form with all its like terms collected is said to be in *standard form*. For example, $2x + 3x + 5$ is an expression written in expanded form; however, to be written in standard form, the like terms $2x$ and $3x$ must be combined. The equivalent expression $5x + 5$ is written in standard form.

Lesson 1: Generating Equivalent Expressions

Lesson Summary

Terms that contain exactly the same variable symbol can be combined by addition or subtraction because the variable represents the same number. Any order, any grouping can be used where terms are added (or subtracted) in order to group together like terms. Changing the orders of the terms in a sum does not affect the value of the expression for given values of the variable(s).

Name _____ Date _____

1. Write an equivalent expression to $2x + 3 + 5x + 6$ by combining like terms.

2. Find the sum of $(8a + 2b - 4)$ and $(3b - 5)$.

3. Write the expression in standard form: $4(2a) + 7(-4b) + (3 \cdot c \cdot 5)$.

1. Write an equivalent expression by combining like terms. Verify the equivalence of your expression and the given expression by evaluating each for the given value: $m = -3$.

   $4m + 7 + m - 9$

   $4m + 1m + 7 - 9$

   $5m - 2$

   > I can rearrange the terms so that I have like terms together. I can also place a 1 in front of the m to make it easier to add $4m + m$.

   > Next, I will replace all of the m's in the original expression and the new expression with -3 and evaluate to see if I get the same result.

   Check:

   $4(-3) + 7 + (-3) - 9$

   $-12 + 7 + (-3) + (-9)$

   -17

   $5(-3) - 2$

   $-15 + (-2)$

   -17

   The expressions $4m + 7 + m - 9$ and $5m - 2$ are equivalent.

2. Use any order and any grouping to write an equivalent expression by combining like terms. Then, verify the equivalence of your expression to the given expression by evaluating for the value(s) given.

   $9(2j) + 6(-7k) + 6(-j)$; for $j = \frac{1}{2}$, $k = \frac{1}{3}$

   $9(2j) + 6(-7k) + 6(-j)$

   $(9)(2)(j) + (6)(-7)(k) + (6)(-1)(j)$

   $18j + (-42k) + (-6j)$

   $18j + (-6j) + (-42k)$

   $12j - 42k$

   > I can multiply in any order, which means I can multiply the 9 and 2 together first for the term $9(2j)$.

Check:

$9(2j) + 6(-7k) + 6(-j)$

$9\left(2 \times \dfrac{1}{2}\right) + 6\left(-7 \times \dfrac{1}{3}\right) + 6\left(-\dfrac{1}{2}\right)$

$9(1) + 6\left(-\dfrac{7}{3}\right) + (-3)$

$9 + \left(-\dfrac{42}{3}\right) + (-3)$

$9 + (-14) + (-3)$

-8

> I evaluate both expressions using the given values for each variable. If I don't get the same result, I might have made an error somewhere in my work.

$12j - 42k$

$12\left(\dfrac{1}{2}\right) + (-42)\left(\dfrac{1}{3}\right)$

$6 + (-14)$

-8

> I can go back to Module 2 for a review on how to work with signed rational numbers.

Both expressions are equivalent.

3. Meredith, Jodi, and Clive were finding the sum of $(5x + 8)$ and $-3x$. Meredith wrote the expression $2x + 8$, Jodi wrote $8x + 2$, and Clive wrote $8 + 2x$. Which person(s) was correct and why?

Let $x = 2$

$(5x + 8) + (-3x)$

$5(2) + 8 + (-3(2))$

$10 + 8 + (-6)$

12

> I could test the equivalence by picking any value for x and evaluating each expression.

Meredith

$2x + 8$

$2(2) + 8$

$4 + 8$

12

> I will replace all the x's in the original expression and the three possible expressions to see if I get the same result.

Lesson 1:   Generating Equivalent Expressions

Jodi

$8x + 2$

$8(2) + 2$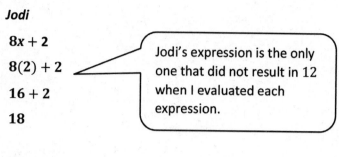

$16 + 2$

18

Jodi's expression is the only one that did not result in 12 when I evaluated each expression.

Clive

$8 + 2x$

$8 + 2(2)$

$8 + 4$

12

Meredith and Clive are correct. Their expressions are the same, just in different orders. Jodi's expression is incorrect.

For Problems 1–9 ,write equivalent expressions by combining like terms. Verify the equivalence of your expression and the given expression by evaluating each for the given values: $a = 2$, $b = 5$, and $c = -3$.

1. $3a + 5a$

2. $8b - 4b$

3. $5c + 4c + c$

4. $3a + 6 + 5a$

5. $8b + 8 - 4b$

6. $5c - 4c + c$

7. $3a + 6 + 5a - 2$

8. $8b + 8 - 4b - 3$

9. $5c - 4c + c - 3c$

Use any order, any grouping to write equivalent expressions by combining like terms. Then, verify the equivalence of your expression to the given expression by evaluating for the value(s) given in each problem.

10. $3(6a)$; for $a = 3$

11. $5d(4)$; for $d = -2$

12. $(5r)(-2)$; for $r = -3$

13. $3b(8) + (-2)(7c)$; for $b = 2$, $c = 3$

14. $-4(3s) + 2(-t)$; for $s = \frac{1}{2}$, $t = -3$

15. $9(4p) - 2(3q) + p$; for $p = -1$, $q = 4$

16. $7(4\,g) + 3(5h) + 2(-3g)$; for $g = \frac{1}{2}$, $h = \frac{1}{3}$

The problems below are follow-up questions to Example 1, part (b) from Classwork: Find the sum of $2x + 1$ and $5x$.

17. Jack got the expression $7x + 1$ and then wrote his answer as $1 + 7x$. Is his answer an equivalent expression? How do you know?

18. Jill also got the expression $7x + 1$, and then wrote her answer as $1x + 7$. Is her expression an equivalent expression? How do you know?

Number Correct: _____

Generating Equivalent Expressions—Round 1

Directions: Write each as an equivalent expression in standard form as quickly and as accurately as possible within the allotted time.

1.	$1 + 1$		23.	$4x + 6x - 12x$	
2.	$1 + 1 + 1$		24.	$4x - 6x + 4x$	
3.	$(1 + 1) + 1$		25.	$7x - 2x + 3$	
4.	$(1 + 1) + (1 + 1)$		26.	$(4x + 3) + x$	
5.	$(1 + 1) + (1 + 1 + 1)$		27.	$(4x + 3) + 2x$	
6.	$x + x$		28.	$(4x + 3) + 3x$	
7.	$x + x + x$		29.	$(4x + 3) + 5x$	
8.	$(x + x) + x$		30.	$(4x + 3) + 6x$	
9.	$(x + x) + (x + x)$		31.	$(11x + 2) - 2$	
10.	$(x + x) + (x + x + x)$		32.	$(11x + 2) - 3$	
11.	$(x + x + x) + (x + x + x)$		33.	$(11x + 2) - 4$	
12.	$2x + x$		34.	$(11x + 2) - 7$	
13.	$3x + x$		35.	$(3x - 9) + (3x + 5)$	
14.	$4x + x$		36.	$(11 - 5x) + (4x + 2)$	
15.	$7x + x$		37.	$(2x + 3y) + (4x + y)$	
16.	$7x + 2x$		38.	$(5x + 1.3y) + (2.9x - 0.6y)$	
17.	$7x + 3x$		39.	$(2.6x - 4.8y) + (6.5x - 1.1y)$	
18.	$10x - x$		40.	$\left(\frac{3}{4}x - \frac{1}{2}y\right) + \left(-\frac{7}{4}x - \frac{5}{2}y\right)$	
19.	$10x - 5x$		41.	$\left(-\frac{2}{5}x - \frac{7}{9}y\right) + \left(-\frac{7}{10}x - \frac{2}{3}y\right)$	
20.	$10x - 10x$		42.	$\left(\frac{1}{2}x - \frac{1}{4}y\right) + \left(-\frac{3}{5}x + \frac{5}{6}y\right)$	
21.	$10x - 11x$		43.	$\left(1.2x - \frac{3}{4}y\right) - \left(-\frac{3}{5}x + 2.25y\right)$	
22.	$10x - 12x$		44.	$\left(3.375x - 8.9y\right) - \left(-7\frac{5}{8}x - 5\frac{2}{5}y\right)$	

Generating Equivalent Expressions—Round 2

Number Correct: _____

Improvement: _____

Directions: Write each as an equivalent expression in standard form as quickly and as accurately as possible within the allotted time.

1.	$1 + 1 + 1$	3
2.	$1 + 1 + 1 + 1$	4
3.	$(1 + 1 + 1) + 1$	4
4.	$(1 + 1 + 1) + (1 + 1)$	5
5.	$(1 + 1 + 1) + (1 + 1 + 1)$	6
6.	$x + x + x$	3x
7.	$x + x + x + x$	4x
8.	$(x + x + x) + x$	4x
9.	$(x + x + x) + (x + x)$	5x
10.	$(x + x + x) + (x + x + x)$	6x
11.	$(x + x + x + x) + (x + x)$	6x
12.	$x + 2x$	3x
13.	$x + 4x$	5x
14.	$x + 6x$	7x
15.	$x + 8x$	9x
16.	$7x + x$	8x
17.	$8x + 2x$	10x
18.	$2x - x$	1x5x
19.	$2x - 2x$	0
20.	$2x - 3x$	-x
21.	$2x - 4x$	-2x
22.	$2x - 8x$	-6x

23.	$3x + 5x - 4x$	4x
24.	$8x - 6x + 4x$	6x
25.	$7x - 4x + 5$	3x+5
26.	$(9x - 1) + x$	10x-1
27.	$(9x - 1) + 2x$	11x-1
28.	$(9x - 1) + 3x$	12x-1
29.	$(9x - 1) + 5x$	14x-1
30.	$(9x - 1) + 6x$	15x-1
31.	$(-3x + 3) - 2$	-3x+1
32.	$(-3x + 3) - 3$	-3x
33.	$(-3x + 3) - 4$	-3x-1
34.	$(-3x + 3) - 5$	-3x-2
35.	$(5x - 2) + (2x + 5)$	7x+3
36.	$(8 - x) + (3x + 2)$	2x+10
37.	$(5x + y) + (x + y)$	6x+2y
38.	$\left(\frac{5}{2}x + \frac{3}{2}y\right) + \left(\frac{11}{2}x - \frac{3}{4}y\right)$	$8x + \frac{3}{4}y$
39.	$\left(\frac{1}{6}x - \frac{3}{8}y\right) + \left(\frac{2}{3}x - \frac{7}{4}y\right)$	$\frac{5}{6}x - \frac{17}{8}y$
40.	$(9.7x - 3.8y) + (-2.8x + 4.5y)$	6.9x+0.7y
41.	$(1.65x - 2.73y) + (-1.35x + 3.76y)$	0.3x+1.03y
42.	$(6.51x - 4.39y) + (-7.46x + 8.11y)$	1.05x+3.72y
43.	$\left(0.7x - \frac{2}{9}y\right) - \left(-\frac{7}{5}x + 2\frac{1}{3}y\right)$	$2.1x - 2\frac{8}{9}y$
44.	$\left(8.4x - 2.25y\right) - \left(-2\frac{1}{2}x - 4\frac{3}{8}y\right)$	

$$\frac{-2}{3} = \frac{21}{9}$$

$$\frac{-2}{9} + \frac{21}{9} = \frac{-23}{9}y$$

Opening Exercise

Additive inverses have a sum of zero. Fill in the center column of the table with the opposite of the given number or expression, then show the proof that they are opposites. The first row is completed for you.

Expression	Opposite	Proof of Opposites
1	-1	$1 + (-1) = 0$
3		
-7		
$-\dfrac{1}{2}$		
x		
$3x$		
$x + 3$		
$3x - 7$		

Example 1: Subtracting Expressions

a. Subtract: $(40 + 9) - (30 + 2)$.

17

b. Subtract: $(3x + 5y - 4) - (4x + 11)$.

$-x + 5y - 15$

Example 2: Combining Expressions Vertically

a. Find the sum by aligning the expressions vertically.

$(5a + 3b - 6c) + (2a - 4b + 13c)$

$$
\begin{array}{r}
5a + 3b - 6c \quad = -6 + 13 = 7\\
+ \; 2a - 4b + 13c\\
\hline
7a + (-b) + 7c\\
7a - b + 7c
\end{array}
$$

b. Find the difference by aligning the expressions vertically.

$(2x + 3y - 4) - (5x + 2)$

$$
\begin{array}{r}
2x + 3y - 4\\
- \; 5x + 0y (-2)\\
\hline
-3x + 3y + (-6)\\
-3x + 3y - 6
\end{array}
$$

Example 3: Using Expressions to Solve Problems

A stick is x meters long. A string is 4 times as long as the stick.

a. Express the length of the string in terms of x.

$$4x$$

b. If the total length of the string and the stick is 15 meters long, how long is the string?

$$3$$

$$15/5 = 3$$

$$x + 4x = 15$$

$$4 \times 3 = 12$$

$$15 - 12 = 3$$

Example 4: Expressions from Word Problems

It costs Margo a processing fee of \$3 to rent a storage unit, plus \$17 per month to keep her belongings in the unit. Her friend Carissa wants to store a box of her belongings in Margo's storage unit and tells her that she will pay her \$1 toward the processing fee and \$3 for every month that she keeps the box in storage. Write an expression in standard form that represents how much Margo will have to pay for the storage unit if Carissa contributes. Then, determine how much Margo will pay if she uses the storage unit for 6 months.

$$(17m + 3) - (3m - 1)$$

$$17m - 3m + 3 + (-1) = 3 + 1$$

$$\boxed{14m + 2}$$

$$m = 6$$

$$\begin{array}{r} 14 \\ \times\ 6 \\ \hline 84 \end{array}$$

$$84 + 2 = \boxed{\$6}$$

Example 5: Extending Use of the Inverse to Division

Multiplicative inverses have a product of 1. Find the multiplicative inverses of the terms in the first column. Show that the given number and its multiplicative inverse have a product of 1. Then, use the inverse to write each corresponding expression in standard form. The first row is completed for you.

Given	Multiplicative Inverse	Proof—Show that their product is 1.	Use each inverse to write its corresponding expression below in standard form.
3	$\dfrac{1}{3}$	$3 \cdot \dfrac{1}{3}$ $\dfrac{3}{1} \cdot \dfrac{1}{3}$ $\dfrac{3}{3}$ $\dfrac{3}{1}$	$12 \div 3$ $12 \cdot \dfrac{1}{3}$ 4
5	$\dfrac{1}{5}$	$5 \cdot \dfrac{1}{5} = \dfrac{5}{5} = 1$	$65 \div 5$
-2	$\dfrac{-1}{2}$	$-2 \times \dfrac{-1}{2} = \dfrac{2}{2} = 1$	$18 \div (-2)$
$-\dfrac{3}{5}$	$-\dfrac{5}{3}$	$\dfrac{-5}{3}$	$6 \div \left(-\dfrac{3}{5}\right)$
x	$\dfrac{1}{x}$		$5x \div x$
$2x$	$\dfrac{1}{2x}$		$12x \div 2x$

Relevant Vocabulary

AN EXPRESSION IN EXPANDED FORM: An expression that is written as sums (and/or differences) of products whose factors are numbers, variables, or variables raised to whole number powers is said to be in *expanded form*. A single number, variable, or a single product of numbers and/or variables is also considered to be in expanded form. Examples of expressions in expanded form include: 324, $3x$, $5x + 3 - 40$, and $x + 2x + 3x$.

TERM: Each summand of an expression in expanded form is called a *term*. For example, the expression $2x + 3x + 5$ consists of 3 terms: $2x$, $3x$, and 5.

COEFFICIENT OF THE TERM: The number found by multiplying just the numbers in a term together is called the *coefficient*. For example, given the product $2 \cdot x \cdot 4$, its equivalent term is $8x$. The number 8 is called the coefficient of the term $8x$.

AN EXPRESSION IN STANDARD FORM: An expression in expanded form with all of its like terms collected is said to be in *standard form*. For example, $2x + 3x + 5$ is an expression written in expanded form; however, to be written in standard form, the like terms $2x$ and $3x$ must be combined. The equivalent expression $5x + 5$ is written in standard form.

Lesson 2:   Generating Equivalent Expressions

Lesson Summary

- Rewrite subtraction as adding the opposite before using any order, any grouping.
- Rewrite division as multiplying by the reciprocal before using any order, any grouping.
- The opposite of a sum is the sum of its opposites.
- Division is equivalent to multiplying by the reciprocal.

Name _____   Date _____

1. Write the expression in standard form.

    $(4f - 3 + 2g) - (-4g + 2)$

2. Find the result when $5m + 2$ is subtracted from $9m$.

3. Write the expression in standard form.

    $27h \div 3h$

1. Write each expression in standard form. Verify that your expression is equivalent to the one given by evaluating both expressions for the given value of the variable.

   a. $5x + (4x - 9); x = 3$

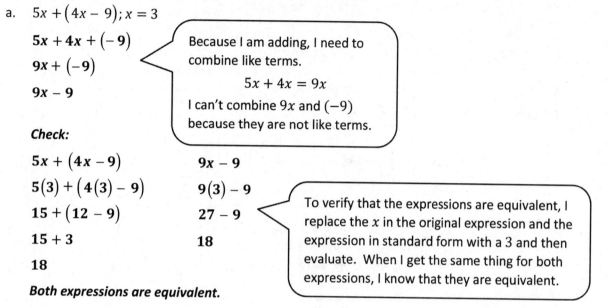

$$5x + 4x + (-9)$$

$$9x + (-9)$$

$$9x - 9$$

Because I am adding, I need to combine like terms.

$$5x + 4x = 9x$$

I can't combine $9x$ and (-9) because they are not like terms.

Check:

$$5x + (4x - 9)$$ $$9x - 9$$

$$5(3) + (4(3) - 9)$$ $$9(3) - 9$$

$$15 + (12 - 9)$$ $$27 - 9$$

$$15 + 3$$ $$18$$

$$18$$

To verify that the expressions are equivalent, I replace the x in the original expression and the expression in standard form with a 3 and then evaluate. When I get the same thing for both expressions, I know that they are equivalent.

Both expressions are equivalent.

   b. $7x - (4 - 2x); x = -5$

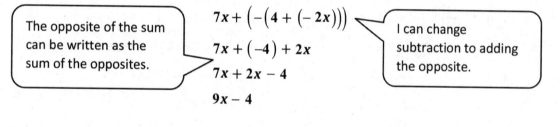

The opposite of the sum can be written as the sum of the opposites.

$$7x + (-(4 + (-2x)))$$

$$7x + (-4) + 2x$$

$$7x + 2x - 4$$

$$9x - 4$$

I can change subtraction to adding the opposite.

Check:

$$7x - (4 - 2x)$$ $$9x - 4$$

$$7(-5) - (4 - 2(-5))$$ $$9(-5) - 4$$

$$-35 - (4 + 10)$$ $$-45 - 4$$

$$-35 - (14)$$ $$-45 + (-4)$$

$$-35 + (-14)$$ $$-49$$

$$-49$$

I can look back to Module 2 for help with order of operations and integers.

These expressions are equivalent.

c. $(11g + 7h - 8) - (3g - 9h + 6); g = -3$ and $h = 4$

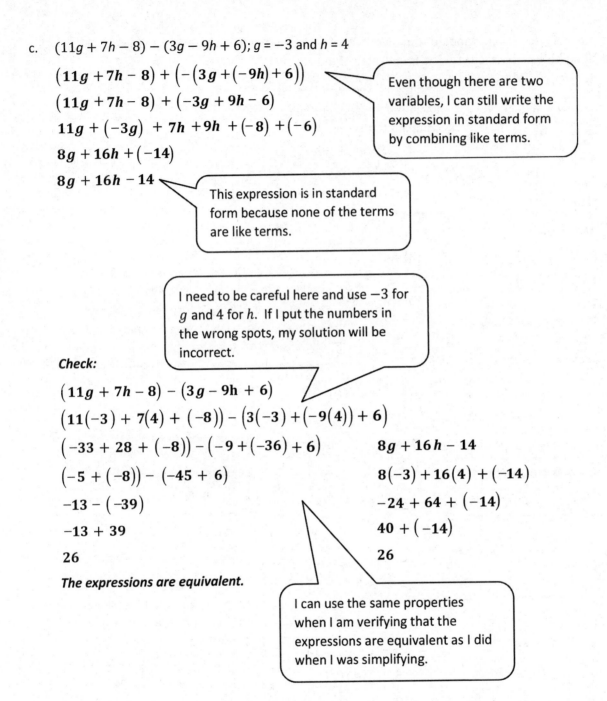

$\left(11g + 7h - 8\right) + \left(-\left(3g + (-9h) + 6\right)\right)$

$\left(11g + 7h - 8\right) + \left(-3g + 9h - 6\right)$

$11g + (-3g) + 7h + 9h + (-8) + (-6)$

$8g + 16h + (-14)$

$8g + 16h - 14$

> Even though there are two variables, I can still write the expression in standard form by combining like terms.

> This expression is in standard form because none of the terms are like terms.

> I need to be careful here and use -3 for g and 4 for h. If I put the numbers in the wrong spots, my solution will be incorrect.

Check:

$\left(11g + 7h - 8\right) - \left(3g - 9h + 6\right)$

$\left(11(-3) + 7(4) + (-8)\right) - \left(3(-3) + (-9(4)) + 6\right)$

$\left(-33 + 28 + (-8)\right) - \left(-9 + (-36) + 6\right)$ $8g + 16h - 14$

$\left(-5 + (-8)\right) - \left(-45 + 6\right)$ $8(-3) + 16(4) + (-14)$

$-13 - \left(-39\right)$ $-24 + 64 + (-14)$

$-13 + 39$ $40 + (-14)$

26 26

The expressions are equivalent.

> I can use the same properties when I am verifying that the expressions are equivalent as I did when I was simplifying.

d. $-3(8v) + 2y(15)$; $v = \frac{1}{4}$; $y = \frac{2}{3}$

$(-3)(8)v + 2(15)$

$-24v + 30y$

> I can multiply in any order. So $2y(15)$ could also be $2(15)y$, giving me $30y$.

Check:

$-3(8v) + 2y(15)$	$-24v + 30y$
$-3\left(8\left(\frac{1}{4}\right)\right) + 2\left(\frac{2}{3}\right)(15)$	$-24\left(\frac{1}{4}\right) + 30\left(\frac{2}{3}\right)$
$-3(2) + 2(10)$	$-6 + 20$
$-6 + 20$	14
14	

The expressions are equivalent.

e. $32xy \div 8y$; $x = -\frac{1}{2}$; $y = 3$

$32xy \times \left(\frac{1}{8y}\right)$

$\dfrac{32xy}{8y}$

$\dfrac{32}{8} \cdot \dfrac{x}{1} \cdot \dfrac{y}{y}$

$4x$

> I can use the "any order, any grouping" property in order to break apart the factors and simplify.

> Dividing is equivalent to multiplying by the reciprocal. So I can rewrite this problem as a multiplication expression.

Check:

$32xy \div 8y$	$4x$
$32\left(-\frac{1}{2}\right)(3) \div 8(3)$	$4\left(-\frac{1}{2}\right)$
$-48 \div 24$	-2
-2	

The expressions are equivalent.

2. Doug and Romel are placing apples in baskets to sell at the farm stand. They are putting x apples in each basket. When they are finished, Doug has 23 full baskets and has 7 extra apples, and Romel has 19 full baskets and has 3 extra apples. Write an expression in standard form that represents the number of apples the boys started with. Explain what your expression means.

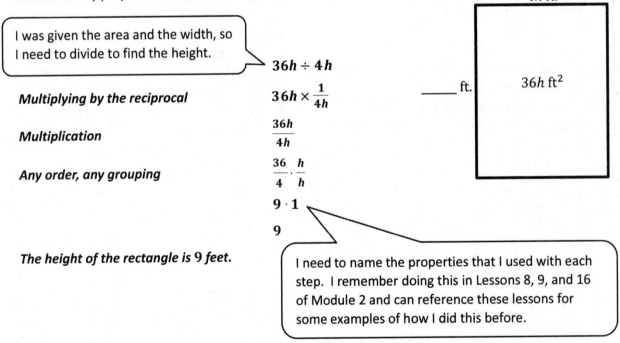

I can represent the number of apples Doug had with the expression $23x + 7$ because he put x apples in 23 baskets.

For Romel, I will use $19x + 3$ because he filled 19 baskets with x apples each. Now I need to add together the number of apples each boy had.

$$23x + 7 + 19x + 3$$
$$23x + 19x + 7 + 3$$
$$42x + 10$$

This means that altogether they have 42 baskets with x apples in each, plus another 10 leftover apples.

3. The area of the pictured rectangle below is $36h$ ft². Its width is $4h$ ft. Find the height of the rectangle, and name any properties used with the appropriate step.

4h ft.

I was given the area and the width, so I need to divide to find the height.

_____ ft. $36h$ ft²

$$36h \div 4h$$

Multiplying by the reciprocal
$$36h \times \frac{1}{4h}$$

Multiplication
$$\frac{36h}{4h}$$

Any order, any grouping
$$\frac{36}{4} \cdot \frac{h}{h}$$

$$9 \cdot 1$$

$$9$$

The height of the rectangle is 9 feet.

I need to name the properties that I used with each step. I remember doing this in Lessons 8, 9, and 16 of Module 2 and can reference these lessons for some examples of how I did this before.

1. Write each expression in standard form. Verify that your expression is equivalent to the one given by evaluating each expression using $x = 5$.

a. $3x + (2 - 4x)$	b. $3x + (-2 + 4x)$	c. $-3x + (2 + 4x)$
$15x - 20x + 2$	$15x + 20x - 2$	$-15x + 20x + 2$
d. $3x + (-2 - 4x)$	e. $3x - (2 + 4x)$	f. $3x - (-2 + 4x)$
$15x - 20x - 2$	$15x - 20x + 2$	$15x - 20x = 2$
g. $3x - (-2 - 4x)$	h. $3x - (2 - 4x)$	i. $-3x - (-2 - 4x)$

     j. In problems (a)–(d) above, what effect does addition have on the terms in parentheses when you removed the parentheses?

     k. In problems (e)–(i), what effect does subtraction have on the terms in parentheses when you removed the parentheses?

2. Write each expression in standard form. Verify that your expression is equivalent to the one given by evaluating each expression for the given value of the variable.

a. $4y - (3 + y)$; $y = 2$	b. $(2b + 1) - b$; $b = -4$	c. $(6c - 4) - (c - 3)$; $c = -7$
d. $(d + 3d) - (-d + 2)$; $d = 3$	e. $(-5x - 4) - (-2 - 5x)$; $x = 3$	f. $11f - (-2f + 2)$; $f = \dfrac{1}{2}$
g. $-5g + (6g - 4)$; $g = -2$	h. $(8h - 1) - (h + 3)$; $h = -3$	i. $(7 + w) - (w + 7)$; $w = -4$
j. $(2g + 9h - 5) - (6g - 4h + 2)$; $g = -2$ and $h = 5$		

3. Write each expression in standard form. Verify that your expression is equivalent to the one given by evaluating both expressions for the given value of the variable.

a. $-3(8x)$; $x = \dfrac{1}{4}$	b. $5 \cdot k \cdot (-7)$; $k = \dfrac{3}{5}$	c. $2(-6x) \cdot 2$; $x = \dfrac{3}{4}$
d. $-3(8x) + 6(4x)$; $x = 2$	e. $8(5m) + 2(3m)$; $m = -2$	f. $-6(2v) + 3a(3)$; $v = \dfrac{1}{3}$; $a = \dfrac{2}{3}$

Lesson 2:   Generating Equivalent Expressions   31

4. Write each expression in standard form. Verify that your expression is equivalent to the one given by evaluating both expressions for the given value of the variable.

a. $8x \div 2$; $x = -\dfrac{1}{4}$	b. $18w \div 6$; $w = 6$	c. $25r \div 5r$; $r = -2$
d. $33y \div 11y$; $y = -2$	e. $56k \div 2k$; $k = 3$	f. $24xy \div 6y$; $x = -2$; $y = 3$

5. For each problem (a)–(g), write an equation in standard form.

   a. Find the sum of $-3x$ and $8x$.

   b. Find the sum of $-7g$ and $4g + 2$.

   c. Find the difference when $6h$ is subtracted from $2h - 4$.

   d. Find the difference when $-3n - 7$ is subtracted from $n + 4$.

   e. Find the result when $13v + 2$ is subtracted from $11 + 5v$.

   f. Find the result when $-18m - 4$ is added to $4m - 14$.

   g. What is the result when $-2x + 9$ is taken away from $-7x + 2$?

6. Marty and Stewart are stuffing envelopes with index cards. They are putting x index cards in each envelope. When they are finished, Marty has 15 stuffed envelopes and 4 extra index cards, and Stewart has 12 stuffed envelopes and 6 extra index cards. Write an expression in standard form that represents the number of index cards the boys started with. Explain what your expression means.

7. The area of the pictured rectangle below is $24b$ ft². Its width is $2b$ ft. Find the height of the rectangle and name any properties used with the appropriate step.

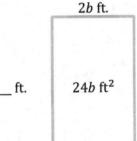

2b ft.

___ ft. $24b$ ft²

Opening Exercise

Solve the problem using a tape diagram. A sum of money was shared between George and Benjamin in a ratio of 3: 4. If the sum of money was $56.00, how much did George get?

Example 1

Represent $3 + 2$ using a tape diagram.

Represent $x + 2$ using a tape diagram.

Draw a rectangular array for $3(3 + 2)$.

Draw an array for $3(x + 2)$.

Key Terms

DISTRIBUTIVE PROPERTY: The *distributive property* can be written as the identity

$$a(b + c) = ab + ac \text{ for all numbers } a, b, \text{ and } c.$$

Exercise 1

Determine the area of each region using the distributive property.

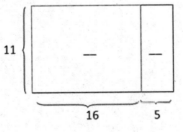

Lesson 3:   Writing Products as Sums and Sums as Products

Example 2

Draw a tape diagram to represent each expression.

   a. $(x + y) + (x + y) + (x + y)$

   b. $(x + x + x) + (y + y + y)$

   c. $3x + 3y$

   d. $3(x + y)$

Example 3

Find an equivalent expression by modeling with a rectangular array and applying the distributive property to the expression $5(8x + 3)$.

$$40x + 15$$

Exercise 2

For parts (a) and (b), draw an array for each expression and apply the distributive property to expand each expression. Substitute the given numerical values to demonstrate equivalency.

    a. $2(x + 1), x = 5$

    b. $10(2c + 5), c = 1$

Example 6

A square fountain area with side length s ft. is bordered by a single row of square tiles as shown. Express the total number of tiles needed in terms of s three different ways.

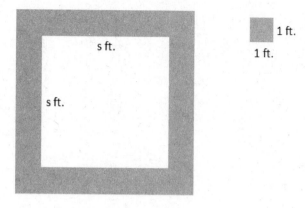

Name _____   Date _____

A square fountain area with side length s ft. is bordered by two rows of square tiles along its perimeter as shown. Express the total number of grey tiles (the second border of tiles) needed in terms of s three different ways.

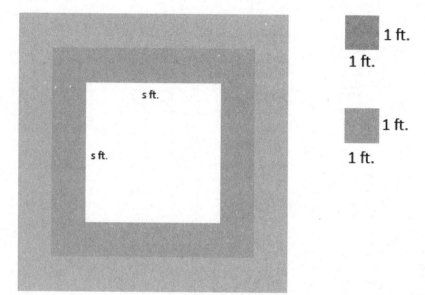

1.

   a. Write two equivalent expressions that represent the rectangular array below.

   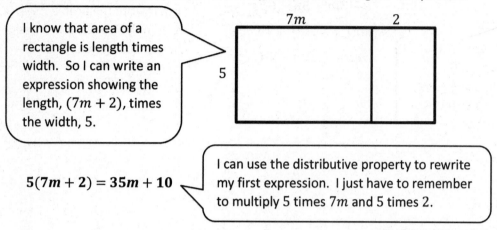

   I know that area of a rectangle is length times width. So I can write an expression showing the length, $(7m + 2)$, times the width, 5.

   $5(7m + 2) = 35m + 10$

   I can use the distributive property to rewrite my first expression. I just have to remember to multiply 5 times $7m$ and 5 times 2.

   b. Verify informally that the two expressions are equivalent using substitution.

   Let $m = 2$

$5(7m + 2)$	$35m + 10$
$5(7(2) + 2)$	$35(2) + 10$
$5(14 + 2)$	$70 + 10$
$5(16)$	80
80	

   To verify that these two expressions are equivalent, I can pick any value I want for m and then substitute it into both expressions to make sure they both give me the same value, just like I did in Lesson 2.

   c. Use a rectangular array to write the product $3(2h + 6g + 4k)$ in standard form.

   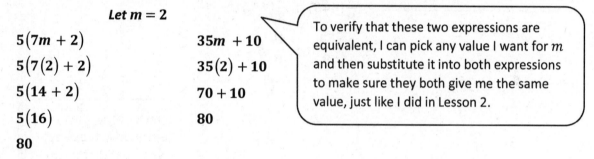

   I draw an array where 3 is the width, and the length is $2h + 6g + 4k$.

   $3(2h) + 3(6g) + 3(4k)$

   $6h + 18g + 12k$

   I multiply each part of the length by the width, 3.

   The expression in standard form is $6h + 18g + 12k$.

2. Use the distributive property to write the products in standard form.

   a. $(3m + 5n - 6p)4$

$$4(3m + 5n - 6p)$$
$$4(3m) + 4(5n) + 4(-6p)$$
$$12m + 20n - 24p$$

> This problem is written a little differently than the others. It is still asking me to distribute the 4 to all terms inside the parentheses.

> I can rewrite the problem with the 4 in the front if that makes it easier for me to simplify.

   b. $(64h + 40g) \div 8$

> I can rewrite the division as multiplication by the reciprocal.

$$\frac{1}{8}(64h + 40g)$$
$$\frac{1}{8}(64h) + \frac{1}{8}(40g)$$
$$\frac{64}{8}h + \frac{40}{8}g$$
$$8h + 5g$$

> I need to use the distributive property twice in this problem.

   c. $7(4x - 1) + 3(5x + 9)$

$$7(4x) + 7(-1) + 3(5x) + 3(9)$$
$$28x + (-7) + 15x + 27$$
$$28x + 15x + 27 + (-7)$$
$$43x + 20$$

> I combine like terms after applying the distributive property.

3. You and your friend are in charge of buying supplies for the next school dance. Each package of balloons costs \$2, and each string of lights costs \$8. Write an equation that represents the total amount spent, S, if b represents the number of packages of balloons purchased and l represents the number of strings of lights purchased. Explain the equation in words.

$$S = 2b + 8l \quad \text{or} \quad S = 2(b + 4l)$$

> I notice that the terms in the first equation have a common factor, which means I can write this equation a second way, by dividing out the common factor from each term and writing it outside the parentheses.

The total amount spent can be determined by multiplying the number of packages of balloons purchased by two and then adding that to the product of the number of strings of lights and eight.

The total amount spent can also be determined by adding the number of packages of balloons purchased to four times the number of strings of lights purchased and then multiplying the sum by two.

1.
   a. Write two equivalent expressions that represent the rectangular array below.

   3(2a + 5)

   b. Verify informally that the two expressions are equivalent using substitution.

2. You and your friend made up a basketball shooting game. Every shot made from the free throw line is worth 3 points, and every shot made from the half-court mark is worth 6 points. Write an equation that represents the total number of points, P, if f represents the number of shots made from the free throw line, and h represents the number of shots made from half-court. Explain the equation in words.

3. Use a rectangular array to write the products in standard form.
   a. $2(x + 10)$
   b. $3(4b + 12c + 11)$

4. Use the distributive property to write the products in standard form.
   a. $3(2x - 1)$
   b. $10(b + 4c)$
   c. $9(g - 5h)$
   d. $7(4n - 5m - 2)$
   e. $a(b + c + 1)$
   f. $(8j - 3l + 9)6$

   g. $(40s + 100t) \div 10$
   h. $(48p + 24) \div 6$
   i. $(2b + 12) \div 2$
   j. $(20r - 8) \div 4$
   k. $(49g - 7) \div 7$
   l. $\left(14g + 22h\right) \div \frac{1}{2}$

5. Write the expression in standard form by expanding and collecting like terms.
   a. $4(8m - 7n) + 6(3n - 4m)$
   b. $9(r - s) + 5(2r - 2s)$
   c. $12(1 - 3g) + 8(g + f)$

Ecample 1

a.	$2(x + 5)$	$2x + 10$
b.	$3(x + 4)$	$3x + 12$
c.	$6(x + 1)$	$6x + 6$
d.	$7(x - 3)$	$7x - 21$
e.	$5(x + 6)$	$5x + 30$
f.	$4(2x + 2)$	$8x + 8$
g.	$3(x - 4)$	$3x - 12$
h.	$3(5x + 4)$	$15x + 20$

Exercise 1

Rewrite the expressions as a product of two factors.

a. $72t + 8$

$8(9 + 1)$

c. $36z + 72$

$9(4z + 8)$

e. $3r + 3s$

$1(3r + 3s)$

b. $55a + 11$

$11(5a + 1)$

d. $144q - 15$

$3(48q - 5)$

Example 2

Let the variables x and y stand for positive integers, and let $2x$, $12y$, and 8 represent the area of three regions in the array. Determine the length and width of each rectangle if the width is the same for each rectangle.

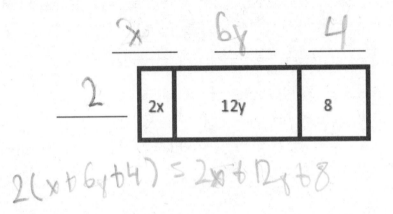

$$2(x + 6y + 4) = 2x + 12y + 8$$

Exercise 2

a. Write the product and sum of the expressions being represented in the rectangular array.

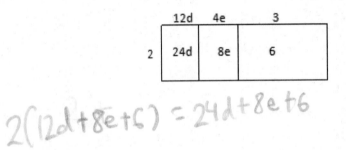

$$2(12d + 8e + 6) = 24d + 8e + 6$$

b. Factor $48j + 60k + 24$ by finding the greatest common factor of the terms.

$$12(4j + 5k + 2)$$

Exercise 3

For each expression, write each sum as a product of two factors. Emphasize the importance of the distributive property. Use various equivalent expressions to justify equivalency.

a. $2 \cdot 3 + 5 \cdot 3$

$3(2+5)$

b. $(2+5) + (2+5) + (2+5)$

$7 \cdot 3$ or

$3(2+5)$

c. $2 \cdot 2 + (5+2) + (5 \cdot 2)$

$2(2+5) + (2+5)^{+5}$

$3(2+5)$

d. $x \cdot 3 + 5 \cdot 3$

$3(x+5)$

e. $(x+5) + (x+5) + (x+5)$

$3(x+5)$

f. $2x + (5+x) + 5 \cdot 2$

$2x + x + 5 + 10$

$3x + 15$

$3(x+5)$

g. $x \cdot 3 + y \cdot 3$

$3(x+y)$

h. $(x+y) + (x+y) + (x+y)$

$3(x+y)$

i. $2x + (y+x) + 2y$

$2x + x + 2y + y$

$3x + 3y$

$3(x+y)$

Example 3

A new miniature golf and arcade opened up in town. For convenient ordering, a play package is available to purchase. It includes two rounds of golf and 20 arcade tokens, plus $3.00 off the regular price. There is a group of six friends purchasing this package. Let g represent the cost of a round of golf, and let t represent the cost of a token. Write two different expressions that represent the total amount this group spent. Explain how each expression describes the situation in a different way.

$g =$ cost of golf round

$t =$ cost of token

$2g =$ rounds

$6(2g + 20t - 3) = 12g + 120t - 18$

cost of each
person by $\times$

total cost for all six people

Exercise 4

a. What is the opposite of $(-6v + 1)$?

$$6v - 1 \qquad -(-6v + 1)$$

b. Using the distributive property, write an equivalent expression for part (a).

$$6v + (-1)$$

Rewrite $5a - (a - 3b)$ in standard form. Justify each step, applying the rules for subtracting and the distributive property.

$$5a - a = 4a$$

$$4a + 3b$$

Exercise 5

Expand each expression and collect like terms.

a. $-3(2p - 3q)$

$$-6p + 9q$$

b. $-a - (a - b)$

Name _____   Date _____

1. Write the expression below in standard form.

    $3h - 2(1 + 4h)$

    $12h \ -2$

2. Write the expression below as a product of two factors.

    $6m + 8n + 4$

    $2(3m + 4n + 2)$

> I need to write my answer showing two expressions that are being multiplied together.

1. Write each expression as the product of two factors.

    a. $k \cdot 5 + m \cdot 5$

        $5(k + m)$

        > I see that both of the addends have a common factor of 5. I can figure out what will still be inside of the parentheses by dividing both terms by 5.

    b. $(d + e) + (d + e) + (d + e) + (d + e)$

        $4(d + e)$

        > I know that repeated addition can be written as multiplication.

    c. $4h + (8 + h) + 3 \cdot 4$

        $4h + 8 + h + 12$

        $5h + 20$

        $5(h + 4)$

        > I must simplify this expression before I can try to write it as the product of two factors.

2. Write each expression in standard form.

    a. $-8(7y - 3z + 5)$

        $-8(7y) + (-8)(-3z) + (-8)(5)$

        $-56y + 24z - 40$

        > To be in standard form, I need to rewrite this expression without the parentheses. I can distribute the -8 to all terms inside.

    b. $4 - 2(-8h - 3)$

        $4 + \left(-2(-8h - 3)\right)$

        $4 + (-2)(-8h) + (-2)(-3)$

        $4 + 16h + 6$

        $10 + 16h$

        > I need to follow the correct order of operations, which means I need to distribute the -2 before I subtract.

© 2019 Great Minds®. eureka-math.org

3. Use the following rectangular array to answer the questions below.

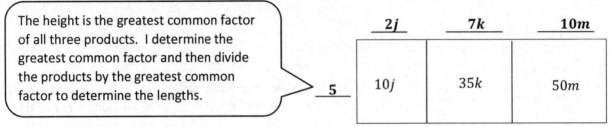

The height is the greatest common factor of all three products. I determine the greatest common factor and then divide the products by the greatest common factor to determine the lengths.

	__2j__	__7k__	__10m__
__5__	$10j$	$35k$	$50m$

a. Fill in the missing information.

b. Write the sum represented in the rectangular array.

$$10j + 35k + 50m$$

I can add the area of each section of the array to write the sum.

c. Use the missing information from part (a) to write the sum from part (b) as a product of two factors.

$$5(2j) + 5(7k) + 5(10m)$$

$$5(2j + 7k + 10m)$$

I need to show that 5 is being multiplied by each le without having to write "times 5" three times.

4. Combine like terms to write each expression in standard form.

$$(-m - n) - (m - n)$$

I know I can rewrite all of the subtraction as adding the opposite.

$$-m + (-n) + (-(m - n))$$

$$-m + (-n) + (-m) + n$$

$$-m + (-m) + (-n) + n$$

$$-2m$$

In the end, I have to add opposites.

$$(-n) + n = 0$$

5. Kathy is a professional dog walker. She must walk the dogs 6 days a week. During each day of walking, she drinks 1 bottle of tea and 3 bottles of water. Let t represent the ounces of tea she drinks and w represent the ounces of water she drinks from each bottle of water. Write two different expressions that represent the total number of ounces Kathy drank in one week while walking the dogs. Explain how each expression describes the situation in a different way.

> In one day, Kathy will drink t ounces of tea and $w + w + w$ or $3w$ ounces of water from the three water bottles. That is $t + 3w$ ounces in one day.

$6(t + 3w)$

Kathy drinks tea and water during walks on six different days, so the total ounces is six times the quantity of the water and tea that Kathy drank each day.

$6(t) + 6(3w)$

$6t + 18w$

There are w bottles of tea and 18 bottles of water total. The total amount that Kathy drank will be six times the ounces in one bottle of tea plus 18 times the ounces in one bottle of water.

1. Write each expression as the product of two factors.

    a. $1 \cdot 3 + 7 \cdot 3$

    b. $(l + 7) + (l + 7) + (l + 7)$

    c. $2 \cdot l + (l + 7) + (7 \cdot 2)$

    d. $h \cdot 3 + 6 \cdot 3$

    e. $(h + 6) + (h + 6) + (h + 6)$

    f. $2h + (6 + h) + 6 \cdot 2$

    g. $j \cdot 3 + k \cdot 3$

    h. $(j + k) + (j + k) + (j + k)$

    i. $2j + (k + j) + 2k$

2. Write each sum as a product of two factors.

    a. $6 \cdot 7 + 3 \cdot 7$

    b. $(8 + 9) + (8 + 9) + (8 + 9)$

    c. $4 + (12 + 4) + (5 \cdot 4)$

    d. $2y \cdot 3 + 4 \cdot 3$

    e. $(x + 5) + (x + 5)$

    f. $3x + (2 + x) + 5 \cdot 2$

    g. $f \cdot 6 + g \cdot 6$

    h. $(c + d) + (c + d) + (c + d) + (c + d)$

    i. $2r + r + s + 2s$

3. Use the following rectangular array to answer the questions below.

?	?	?	?
?	15f	5g	45

    a. Fill in the missing information.

    b. Write the sum represented in the rectangular array.

    c. Use the missing information from part (a) to write the sum from part (b) as a product of two factors.

4. Write the sum as a product of two factors.

    a. $81w + 48$

    b. $10 - 25t$

    c. $12a + 16b + 8$

5. Xander goes to the movies with his family. Each family member buys a ticket and two boxes of popcorn. If there are five members of his family, let t represent the cost of a ticket and p represent the cost of a box of popcorn. Write two different expressions that represent the total amount his family spent. Explain how each expression describes the situation in a different way.

6. Write each expression in standard form.

    a. $-3(1 - 8m - 2n)$ $-3 + 24m + 6n$

    b. $5 - 7(-4q + 5)$ $8q = 10$

    c. $-(2h - 9) - 4h$ $2h + (-9) + (-4h) = -2h - 9$

    d. $6(-5r - 4) - 2(r - 7s - 3)$

7. Combine like terms to write each expression in standard form.

    a. $(r - s) + (s - r)$

    b. $(-r + s) + (s - r)$

    c. $(-r - s) - (-s - r)$

    d. $(r - s) + (s - t) + (t - r)$

    e. $(r - s) - (s - t) - (t - r)$

Lesson 4:   Writing Products as Sums and Sums as Products

© 2019 Great Minds®. eureka-math.org

Opening Exercise

a. In the morning, Harrison checked the temperature outside to find that it was −12°F. Later in the afternoon, the temperature rose 12°F. Write an expression representing the temperature change. What was the afternoon temperature?

$$-12 + 12 = \boxed{0.°F}$$

b. Rewrite subtraction as adding the inverse for the following problems and find the sum.

   i. $2 - 2$

$$2 + (-2) = 0$$

   ii. $-4 - (-4)$

$$-4 + 4 = 0$$

   iii.   The difference of 5 and 5

$$5 + (-5) = 0$$

   iv. $g - g$

$$g + (-g) = 0$$

c.   What pattern do you notice in part (a) and (b)?

Sum of number and addition inverse is = to 0.

d.   Add or subtract.

i. $16 + 0$ = 16

ii. $0 - 7$ = -7

iii. $-4 + 0$ = -4

iv. $0 + d$ = d

v.   What pattern do you notice in parts (i) through (iv)?

equal to valued number

e.   Your younger sibling runs up to you and excitedly exclaims, "I'm thinking of a number. If I add it to the number 2 ten times, that is, 2 + my number + my number + my number, and so on, then the answer is 2. What is my number?" You almost immediately answer, "zero," but are you sure? Can you find a different number (other than zero) that has the same property? If not, can you justify that your answer is the only correct answer?

No other answer than 0.

Lesson 5:   Using the Identity and Inverse to Write Equivalent
                          Expressions

© 2019 Great Minds®. eureka-math.org

Example 1

Write the sum, and then write an equivalent expression by collecting like terms and removing parentheses.

   a. $2x$ and $-2x + 3$

$$2x + (-2x + 3)$$
$$2x + (-2x) + 3$$
$$x + 3 \quad 0 + 3 = 3$$

$$0x = 0$$

   b. $2x - 7$ and the opposite of $2x$

$$2x - 7 + (-2x)$$
$$2x + (-2x) + 7$$
$$x + 7 \quad -0 + 7 = 7$$

   c.   The opposite of $(5x - 1)$ and $5x$

$$-5x - (-1) + 5x \rightarrow -5x + 1 + 5x$$
$$-5x + 5x = 0 \, x \, or \, x$$
$$0 + 1 = 1$$

Exercise 1

With a partner, take turns alternating roles as writer and speaker. The speaker verbalizes how to rewrite the sum and properties that justify each step as the writer writes what is being spoken without any input. At the end of each problem, discuss in pairs the resulting equivalent expressions.

Write the sum, and then write an equivalent expression by collecting like terms and removing parentheses whenever possible.

   a. -4 and $4b + 4$

   b. $3x$ and $1 - 3x$

c. The opposite of $4x$ and $-5 + 4x$

d. The opposite of $-10t$ and $t - 10t$

e. The opposite of $(-7 - 4v)$ and $-4v$

Example 2

- $\left(\dfrac{3}{4}\right) \times \left(\dfrac{4}{3}\right) =$ $\dfrac{12}{12} = 1$
- $4 \times \dfrac{1}{4} =$ $\dfrac{4}{4} = 1$
- $\dfrac{1}{9} \times 9 =$ $\dfrac{9}{9} = 1$
- $\left(-\dfrac{1}{3}\right) \times -3 =$ $\dfrac{3}{3} = 1$
- $\left(-\dfrac{6}{5}\right) \times \left(-\dfrac{5}{6}\right) =$ $\dfrac{30}{30} = 1$

Write the product, and then write the expression in standard form by removing parentheses and combining like terms. Justify each step.

a. The multiplicative inverse of $\dfrac{1}{5}$ and $\left(2x - \dfrac{1}{5}\right)$

$5\left(2x - \dfrac{1}{5}\right)$

$10x - 1$

b. The multiplicative inverse of 2 and $(2x + 4)$

$\dfrac{1}{2}(2x + 4)$

$x + 2$

Lesson 5:   Using the Identity and Inverse to Write Equivalent
                              Expressions

© 2019 Great Minds®. eureka-math.org

c.   The multiplicative inverse of $\left(\frac{1}{3x+5}\right)$ and $\frac{1}{3}$

$$\frac{1}{3} \cdot \left(3x+5\right)\frac{1}{3}$$

$$x \quad + \frac{5}{3}$$

Exercise 2

Write the product, and then write the expression in standard form by removing parentheses and combining like terms. Justify each step.

a.   The reciprocal of 3 and $-6y - 3x$

b.   The multiplicative inverse of 4 and $4h - 20$

c.   The multiplicative inverse of $-\frac{1}{6}$ and $2 - \frac{1}{6}j$

Name _____ Date _____

1. Find the sum of $5x + 20$ and the opposite of 20. Write an equivalent expression in standard form. Justify each step.

2. For $5x + 20$ and the multiplicative inverse of 5, write the product and then write the expression in standard form, if possible. Justify each step.

1. Fill in the missing parts.

   The product of $\frac{1}{3}g + 4$ and the multiplicative inverse of $\frac{1}{3}$

   > The first part has been set up for me, and it shows that 3 is the multiplicative inverse of $\frac{1}{3}$.

   > I see that the column on the left shows the steps, and the column on the right shows the properties that describe the steps.

   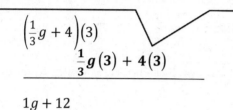
   $$\left(\frac{1}{3}g + 4\right)(3)$$
   $$\frac{1}{3}g(3) + 4(3)$$

   Distributive property

   $$1g + 12$$

   Multiplicative inverse; multiplication

   $$g + 12$$

   Multiplicative identity property of one

   > Here, I can rewrite the expression without the 1 because g and $1g$ are equivalent expressions.

2. Write the sum, and then rewrite the expression in standard form by removing parentheses and collecting like terms.

   a. 13 and $4w - 13$

   $$13 + (4w - 13)$$
   $$13 + 4w + (-13)$$
   $$13 + (-13) + 4w$$
   $$4w$$

   > I can rewrite subtraction as adding the opposite so that all terms are being added.

   > I use the additive inverse property, showing that a number and its inverse have a sum of 0.

   b. The opposite of $5m$ and $9 + 5m$

   $$-5m + (9 + 5m)$$
   $$-5m + 5m + 9$$
   $$9$$

   > Because this question says "the opposite of $5m$," I use the opposite sign, making the term $-5m$.

c. $7y$ and the opposite of $(3 - 8y)$

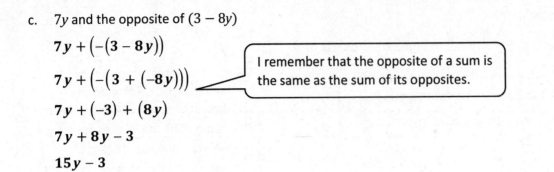

$$7y + \left(-(3 - 8y)\right)$$

$$7y + \left(-(3 + (-8y))\right)$$

I remember that the opposite of a sum is the same as the sum of its opposites.

$$7y + (-3) + (8y)$$

$$7y + 8y - 3$$

$$15y - 3$$

3. Write the product, and then rewrite the expression in standard form by removing parentheses and collecting like terms.

The multiplicative inverse of -8 and $24g - 8$

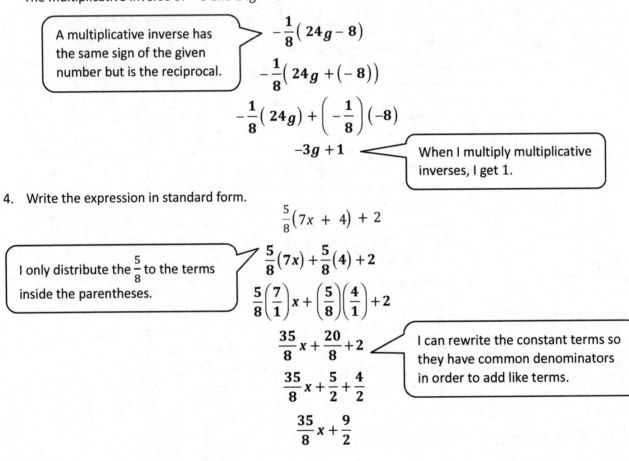

A multiplicative inverse has the same sign of the given number but is the reciprocal.

$$-\frac{1}{8}(24g - 8)$$

$$-\frac{1}{8}(24g + (-8))$$

$$-\frac{1}{8}(24g) + \left(-\frac{1}{8}\right)(-8)$$

$$-3g + 1$$

When I multiply multiplicative inverses, I get 1.

4. Write the expression in standard form.

$$\frac{5}{8}(7x + 4) + 2$$

I only distribute the $\frac{5}{8}$ to the terms inside the parentheses.

$$\frac{5}{8}(7x) + \frac{5}{8}(4) + 2$$

$$\frac{5}{8}\left(\frac{7}{1}\right)x + \left(\frac{5}{8}\right)\left(\frac{4}{1}\right) + 2$$

$$\frac{35}{8}x + \frac{20}{8} + 2$$

I can rewrite the constant terms so they have common denominators in order to add like terms.

$$\frac{35}{8}x + \frac{5}{2} + \frac{4}{2}$$

$$\frac{35}{8}x + \frac{9}{2}$$

Lesson 5: Using the Identity and Inverse to Write Equivalent Expressions

EUREKA MATH

1. Fill in the missing parts.

a.   The sum of $6c - 5$ and the opposite of $6c$

$$\left(6c - 5\right) + \left(-6c\right)$$

_____   Rewrite subtraction as addition

$6c + \left(-6c\right) + \left(-5\right)$ _____

$0 + \left(-5\right)$ _____

_____   Additive identity property of zero

b.   The product of $-2c + 14$ and the multiplicative inverse of -2

$$\left(-2c + 14\right)\left(-\tfrac{1}{2}\right)$$

$$\left(-2c\right)\left(-\tfrac{1}{2}\right) + \left(14\right)\left(-\tfrac{1}{2}\right)$$ _____

_____   Multiplicative inverse, multiplication

$1c - 7$ Adding the additive inverse is the same as subtraction

$c - 7$ _____

2. Write the sum, and then rewrite the expression in standard form by removing parentheses and collecting like terms.

a. 6 and $p - 6$

b. $10w + 3$ and -3

c. $-x - 11$ and the opposite of -11

d.   The opposite of $4x$ and $3 + 4x$

e. $2g$ and the opposite of $(1 - 2g)$

3. Write the product, and then rewrite the expression in standard form by removing parentheses and collecting like terms.

a. $7h - 1$ and the multiplicative inverse of 7

b.   The multiplicative inverse of -5 and $10v - 5$

c. $9 - b$ and the multiplicative inverse of 9

d.   The multiplicative inverse of $\tfrac{1}{4}$ and $5t - \tfrac{1}{4}$

e.   The multiplicative inverse of $-\dfrac{1}{10x}$ and $\dfrac{1}{10x} - \dfrac{1}{10}$

4. Write the expressions in standard form.

   a. $\frac{1}{4}(4x + 8)$ $1x + 8$

   b. $\frac{1}{6}(r - 6)$ $\frac{r}{6} - 1$

   c. $\frac{4}{5}(x + 1)$ $\frac{4}{5}x + \frac{4}{5}$

   d. $\frac{1}{8}(2x + 4)$ $\frac{1}{4}x + \frac{1}{2}$

   e. $\frac{3}{4}(5x - 1)$ $\frac{15}{4}x - \frac{3}{4}$

   f. $\frac{1}{5}(10x - 5) - 3$ $2x - 1 - 3$

Lesson 5:   Using the Identity and Inverse to Write Equivalent
                              Expressions

EUREKA
MATH

Opening Exercise

Solve each problem, leaving your answers in standard form. Show your steps.

a.   Terry weighs 40 kg. Janice weighs $2\frac{3}{4}$ kg less than Terry. What is their combined weight?

$$40 + \left(40 \cdot 2\frac{3}{4}\right) = 77\frac{1}{4} \text{ kg}$$

$$40 + 37\frac{1}{4} = 77\frac{1}{4}$$

b. $2\frac{2}{3} - 1\frac{1}{2} - \frac{4}{5}$

$$\frac{8}{3} - \frac{3}{2} = \frac{16}{6} - \frac{9}{6} = \frac{7}{6} = 1\frac{1}{6}, \quad \frac{7}{6} - \frac{4}{5} = \frac{35}{30} - \frac{24}{30} = \boxed{\frac{11}{30}}$$

c. $\frac{1}{5} + (-4)$

$$\frac{1}{5} - \frac{4}{1} = \frac{1}{5} - \frac{20}{5} = \boxed{\frac{-19}{5}}$$

d. $4\left(\frac{3}{5}\right)$

$$\frac{4}{1} \cdot \frac{3}{5} \quad \boxed{\frac{12}{5}}$$

e.   Mr. Jackson bought $1\frac{3}{5}$ lb. of beef. He cooked $\frac{3}{4}$ of it for lunch. How much does he have left?

$$\frac{8}{5} \cdot \frac{3}{4} = \frac{24}{20} = 1\frac{1}{5} \text{ pond left}$$

Example 1

Rewrite the expression in standard form by collecting like terms.

$$\frac{2}{3}n - \frac{3}{4}n + \frac{1}{6}n + 2\frac{2}{9}n$$

$$\frac{24}{36}n - \frac{27}{36}n + \frac{6}{36}n + \frac{80}{36}n = \frac{83}{36} = 2\frac{11}{36}n$$

Exercise 1

For the following exercises, predict how many terms the resulting expression will have after collecting like terms. Then, write the expression in standard form by collecting like terms.

a. $\frac{2}{5}g - \frac{1}{6} - g + \frac{3}{10}g - \frac{4}{5}$

b. $i + 6i - \frac{3}{7}i + \frac{1}{3}h + \frac{1}{2}i - h + \frac{1}{4}h$

Example 2

At a store, a shirt was marked down in price by $10.00. A pair of pants doubled in price. Following these changes, the price of every item in the store was cut in half. Write two different expressions that represent the new cost of the items, using s for the cost of each shirt and p for the cost of a pair of pants. Explain the different information each one shows.

$$\frac{1}{2}(s - 10) \qquad \frac{1}{2}(2p)$$

$$\frac{1}{2}s - 5 \qquad p$$

Exercise 2

Write two different expressions that represent the total cost of the items if tax was $\frac{1}{10}$ of the original price. Explain the different information each shows.

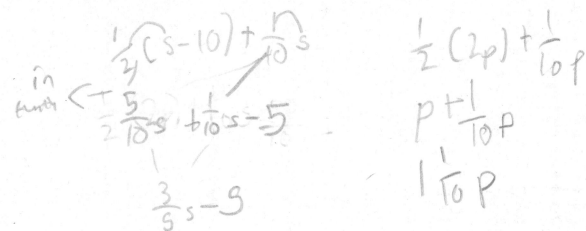

in
terms $\left\{ +\frac{1}{2}(s-10) + \frac{n}{10}s \right.$

$\frac{5}{10}s + \frac{1}{10}s - 5$

$\frac{3}{5}s - 5$

$\frac{1}{2}(2p) + \frac{1}{10}p$

$p + \frac{1}{10}p$

$1\frac{1}{10}p$

Example 3

Write this expression in standard form by collecting like terms. Justify each step.

$$5\frac{1}{3} - \left(3\frac{1}{3}\right)\left(\frac{1}{2}x - \frac{1}{4}\right)$$

$5\frac{1}{3} + \left(-\frac{10}{3}\right)\left(\frac{1}{2}x - \frac{1}{4}\right)$

$\frac{16}{3} + \left(-\frac{5}{3}x\right) + \frac{5}{6}$

$\frac{32}{6} + \frac{5}{6} = \frac{37}{6}$

becomes
positive
after neg
multiply

$\frac{37}{6} + \frac{5}{3}x$

Exercise 3

Rewrite the following expressions in standard form by finding the product and collecting like terms.

a. $-6\frac{1}{3}-\frac{1}{2}\left(\frac{1}{2}+y\right)$

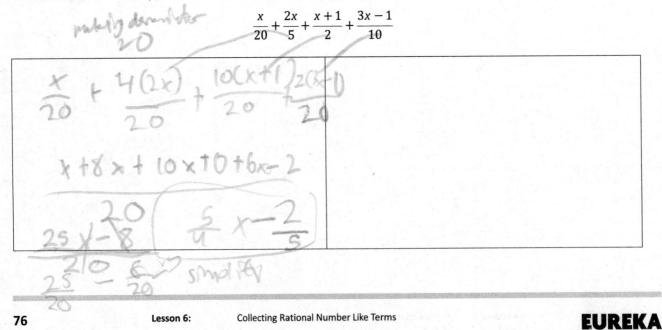

$-\dfrac{19}{3}+\left(-\dfrac{1}{2}\right)\left(\dfrac{1}{2}+y\right)$

$-\dfrac{19}{3}+\dfrac{1}{4}+\dfrac{1}{2}y$ $-\dfrac{79}{12}+\frac{1}{2}y$

$-\dfrac{76}{12}+\dfrac{3}{6}=-\dfrac{79}{12}$

b. $\frac{2}{3}+\frac{1}{3}\left(\frac{1}{4}f-1\frac{1}{3}\right)$

Example 4

Model how to write the expression in standard form using rules of rational numbers.

$$\frac{x}{20}+\frac{2x}{5}+\frac{x+1}{2}+\frac{3x-1}{10}$$

make a denominator 20

$\dfrac{x}{20}+\dfrac{4(2x)}{20}+\dfrac{10(x+1)}{20}+\dfrac{2(x-1)}{20}$

$\dfrac{x+8x+10x+10+6x-2}{20}$

$\dfrac{25x-8}{20}$ $\dfrac{5}{4}x-\dfrac{2}{5}$

$\dfrac{25}{20}=\dfrac{6}{20}$ simplify

EUREKA MATH

Evaluate the original expression and the answers when $x = 20$. Do you get the same number?

Exercise 4

Rewrite the following expression in standard form by finding common denominators and collecting like terms.

$$\frac{2h}{3} - \frac{h}{9} + \frac{h-4}{6}$$

Example 5

Rewrite the following expression in standard form.

$$\frac{2(3x-4)}{6} - \frac{5x+2}{8}$$

Method 1:	Method 2a:	Method 2b:	Method 3:

Exercise 5

Write the following expression in standard form.

$$\frac{2x-11}{4} - \frac{3(x-2)}{10}$$

Name _____ Date _____

For the problem $\frac{1}{5}g - \frac{1}{10} - g + 1\frac{3}{10}g - \frac{1}{10}$, Tyson created an equivalent expression using the following steps.

$$\frac{1}{5}g + -1g + 1\frac{3}{10}g + -\frac{1}{10} + -\frac{1}{10}$$

$$-\frac{4}{5}g + 1\frac{1}{10}$$

Is his final expression equivalent to the initial expression? Show how you know. If the two expressions are not equivalent, find Tyson's mistake and correct it.

1. Write the indicated expression.

a. $\frac{2}{5}k$ inches in yards

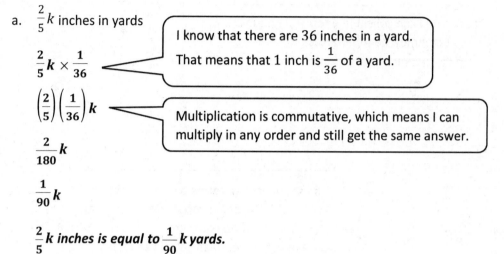

$$\frac{2}{5}k \times \frac{1}{36}$$

> I know that there are 36 inches in a yard.
> That means that 1 inch is $\frac{1}{36}$ of a yard.

$$\left(\frac{2}{5}\right)\left(\frac{1}{36}\right)k$$

> Multiplication is commutative, which means I can multiply in any order and still get the same answer.

$$\frac{2}{180}k$$

$$\frac{1}{90}k$$

$\frac{2}{5}k$ inches is equal to $\frac{1}{90}k$ yards.

b. The average speed of a bike rider that travels $3m$ miles in $\frac{5}{8}$ hour

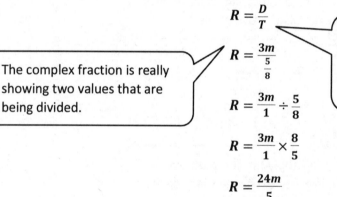

$$R = \frac{D}{T}$$

$$R = \frac{3m}{\frac{5}{8}}$$

> I know that distance is equal to the rate multiplied by the time. I can write this formula so that I am solving for the rate instead.

> The complex fraction is really showing two values that are being divided.

$$R = \frac{3m}{1} \div \frac{5}{8}$$

$$R = \frac{3m}{1} \times \frac{8}{5}$$

$$R = \frac{24m}{5}$$

The average speed of the bike rider is $\frac{24m}{5}$ miles per hour.

Lesson 6:   Collecting Rational Number Like Terms

81

© 2019 Great Minds®. eureka-math.org

2. Rewrite the expression by collecting like terms.

a. $\dfrac{b}{5} - \dfrac{3b}{4} + 2$

> Before I can collect like terms, I need to get common denominators.

$\dfrac{4b}{20} - \dfrac{15b}{20} + 2$

$\dfrac{4b}{20} + \left(-\dfrac{15b}{20}\right) + 2$

$-\dfrac{11b}{20} + 2$

> I must collect like terms by combining the terms with the same variable part. To do this, I need to find common denominators for each set of like terms.

b. $\dfrac{2}{3}k - k - \dfrac{5}{6}k + \dfrac{4}{5} - \dfrac{3}{5}m + 3\dfrac{1}{10}m$

$\dfrac{4}{6}k - \dfrac{6}{6}k - \dfrac{5}{6}k + \dfrac{4}{5} - \dfrac{6}{10}m + \dfrac{31}{10}m$

$-\dfrac{7}{6}k + \dfrac{25}{10}m + \dfrac{4}{5}$

> Before I can collect like terms, I must apply the distributive property.

c. $\dfrac{2}{3}(g + 5) - \dfrac{1}{4}(8g + 1)$

$\dfrac{2}{3}(g) + \dfrac{2}{3}(5) + \left(-\dfrac{1}{4}\right)(8g) + \left(-\dfrac{1}{4}\right)(1)$

$\dfrac{2}{3}g + \dfrac{10}{3} + (-2g) + \left(-\dfrac{1}{4}\right)$

> I can apply the commutative property to change the order so that the like terms are together.

$\dfrac{2}{3}g + (-2g) + \dfrac{10}{3} + \left(-\dfrac{1}{4}\right)$

$\dfrac{2}{3}g + \left(-\dfrac{6}{3}g\right) + \dfrac{40}{12} + \left(-\dfrac{3}{12}\right)$

$-\dfrac{4}{3}g + \dfrac{37}{12}$

d. $\dfrac{5y}{3} + \dfrac{2y+1}{4} - \dfrac{y-7}{2}$

$\dfrac{5y}{3} + \dfrac{2y+1}{4} + \left(-\left(\dfrac{y-7}{2}\right)\right)$

> I remember that the opposite of a sum is the same as the sum of its opposites.

$\dfrac{5y}{3} + \dfrac{2y+1}{4} + \dfrac{-y+7}{2}$

$\dfrac{4(5y)}{4(3)} + \dfrac{3(2y+1)}{3(4)} + \dfrac{6(-y+7)}{6(2)}$

> Getting common denominators will make it easier to collect the like terms in the numerator.

$\dfrac{20y}{12} + \dfrac{6y+3}{12} + \dfrac{-6y+42}{12}$

$\dfrac{20y+6y-6y+3+42}{12}$

$\dfrac{20y+45}{12}$

> Or, I could write my answer as $\dfrac{5y}{3} + \dfrac{15}{4}$.

1. Write the indicated expressions.

    a. $\frac{1}{2}m$ inches in feet.

    b. The perimeter of a square with $\frac{2}{3}g$ cm sides.

    c. The number of pounds in 9 oz.

    d. The average speed of a train that travels x miles in $\frac{3}{4}$ hour.

    e. Devin is $1\frac{1}{4}$ years younger than Eli. April is $\frac{1}{5}$ as old as Devin. Jill is 5 years older than April. If Eli is E years old, what is Jill's age in terms of E?

2. Rewrite the expressions by collecting like terms.

    a. $\frac{1}{2}k - \frac{3}{8}k$

    b. $\frac{2r}{5} + \frac{7r}{15}$

    c. $-\frac{1}{3}a - \frac{1}{2}b - \frac{3}{4} + \frac{1}{2}b + \frac{2}{3}b + \frac{5}{6}a$

    d. $-p + \frac{3}{5}q - \frac{1}{10}q + \frac{1}{9} - \frac{1}{9}p + 2\frac{1}{3}p$

    e. $\frac{5}{7}y - \frac{y}{14}$

    f. $\frac{3n}{8} - \frac{n}{4} + 2\frac{n}{2}$

3. Rewrite the expressions by using the distributive property and collecting like terms.

    a. $\frac{4}{5}(15x - 5)$

    b. $\frac{4}{5}\left(\frac{1}{4}c - 5\right)$

    c. $2\frac{4}{5}v - \frac{2}{3}\left(4v + 1\frac{1}{6}\right)$

    d. $8 - 4\left(\frac{1}{8}r - 3\frac{1}{2}\right)$

    e. $\frac{1}{7}(14x + 7) - 5$

    f. $\frac{1}{5}(5x - 15) - 2x$

    g. $\frac{1}{4}(p + 4) + \frac{3}{5}(p - 1)$

    h. $\frac{7}{8}(w + 1) + \frac{5}{6}(w - 3)$

    i. $\frac{4}{5}(c - 1) - \frac{1}{8}(2c + 1)$

    j. $\frac{2}{3}\left(h + \frac{3}{4}\right) - \frac{1}{3}\left(h + \frac{3}{4}\right)$

    k. $\frac{2}{3}\left(h + \frac{3}{4}\right) - \frac{2}{3}\left(h - \frac{3}{4}\right)$

    l. $\frac{2}{3}\left(h + \frac{3}{4}\right) + \frac{2}{3}\left(h - \frac{3}{4}\right)$

    m. $\frac{k}{2} - \frac{4k}{5} - 3$

    n. $\frac{3t + 2}{7} + \frac{t - 4}{14}$

    o. $\frac{9x - 4}{10} + \frac{3x + 2}{5}$

    p. $\frac{3(5g - 1)}{4} - \frac{2g + 7}{6}$

    q. $-\frac{3d + 1}{5} + \frac{d - 5}{2} + \frac{7}{10}$

    r. $\frac{9w}{6} + \frac{2w - 7}{3} - \frac{w - 5}{4}$

    s. $\frac{1 + f}{5} - \frac{1 + f}{3} + \frac{3 - f}{6}$

Opening Exercise

Your brother is going away to college, so you no longer have to share a bedroom. You decide to redecorate a wall by hanging two new posters on the wall. The wall is 14 feet wide and each poster is four feet wide. You want to place the posters on the wall so that the distance from the edge of each poster to the nearest edge of the wall is the same as the distance between the posters, as shown in the diagram below. Determine that distance.

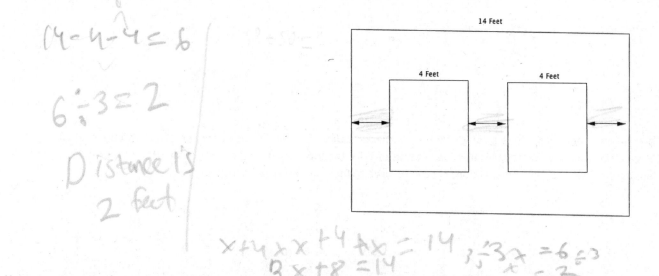

$14 - 4 - 4 = 6$

$6 \div 3 = 2$

Distance is 2 feet

$x + 4 + x + 4 + x = 14$
$3x + 8 = 14$ $3x = 6$
         $x = 2$

Your parents are redecorating the dining room and want to place two rectangular wall sconce lights that are 25 inches wide along a $10\frac{2}{3}$-foot wall so that the distance between the lights and the distances from each light to the nearest edge of the wall are all the same. Design the wall and determine the distance.

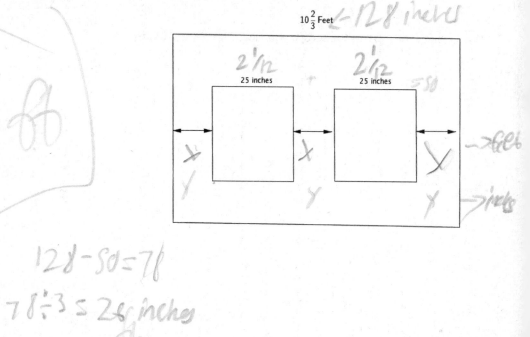

$128 - 50 = 78$

$78 \div 3 = 26$ inches

Let the distance between a light and the nearest edge of a wall be x ft. Write an expression in terms of x for the total length of the wall. Then, use the expression and the length of the wall given in the problem to write an equation that can be used to find that distance.

$$x + 2\tfrac{1}{12} + x + 2\tfrac{1}{12} + x = 10\tfrac{3}{12} \text{ or } 10\tfrac{1}{3}$$

$$3x + 4\tfrac{2}{12} = 10\tfrac{3}{12}$$
$$\qquad -4\tfrac{2}{12}$$
$$3x = 6\tfrac{1}{2} \div 3$$
$$x = 2\tfrac{1}{6}$$

Now write an equation where y stands for the number of *inches*: Let the distance between a light and the nearest edge of a wall be y inches. Write an expression in terms of y for the total length of the wall. Then, use the expression and the length of the wall to write an equation that can be used to find that distance (in inches).

$$3y + 50 = 128$$
$$\quad -50 \quad -50$$
$$3 \div 3y = 78 \div 3$$
$$\boxed{y = 26}$$

$$10 \times 12 = 120 \text{ in}$$
$$2/3 \times 12 = 8 \text{ in}$$
$$128 \text{ in}$$
$$w = 128 \text{ in}$$

What value(s) of y makes the second equation true: 24, 25, or 26?

$$3y + 50 = 128$$

$$3(24) + 50 = 128$$
$$72 + 50 = 122 \quad F$$

$$3(25) + 50 = 128$$
$$75 + 50 = 125 \quad F$$

$$3(26) + 50 = 128$$
$$78 + 50 = 128 \quad T$$

equation is true
when y is 26.

Example

The ages of three sisters are consecutive integers. The sum of their ages is 45. Calculate their ages.

a. Use a tape diagram to find their ages.

youngest [X   ]

middle   [X   ][1]

oldest   [X   ][1][1]

} 45

$45 - 3 = 42$

$42 \div 3 = 14$

youngest = 14

middle = 15

oldest = 16

b. If the youngest sister is x years old, describe the ages of the other two sisters in terms of x, write an expression for the sum of their ages in terms of x, and use that expression to write an equation that can be used to find their ages.

youngest x
middle $x + 1$
oldest $x + (1 + 1)$ or $x + 2$

Sum of ages = $x + (x + 1) + (x + 2) = 45$

$3x + 3 = 45$
$3x = 42$
$x = 14$

$y = 14$
$m = 15$
$o = 16$

c. Determine if your answer from part (a) is a solution to the equation you wrote in part (b).

$14 + (14 + 1) + (14 + 2) = 45$
$14 + (15) + (16) = 45$

Exercise

Sophia pays a $19.99 membership fee for an online music store.

a. If she also buys two songs from a new album at a price of $0.99 each, what is the total cost?

$x = $ total cost

$0.99 \cdot 2 \leq x$

$= 1.98 + 19.99 \leq 21.97$

b. If Sophia purchases n songs for $0.99 each, write an expression for the total cost.

$19.99 + 0.99n = x$

c. Sophia's friend has saved $118 but is not sure how many songs she can afford if she buys the membership and some songs. Use the expression in part (b) to write an equation that can be used to determine how many songs Sophia's friend can buy. $s = $ songs

$\$9 + 19.99 = 118$

-19.99

$s = 99.01$

d. Using the equation written in part (c), can Sophia's friend buy 101, 100, or 99 songs?

No, 98

Relevant Vocabulary

VARIABLE (DESCRIPTION): A *variable* is a symbol (such as a letter) that represents a number (i.e., it is a placeholder for a number).

EQUATION: An *equation* is a statement of equality between two expressions.

NUMBER SENTENCE: A *number sentence* is a statement of equality between two numerical expressions.

SOLUTION: A *solution* to an equation with one variable is a number that, when substituted for the variable in both expressions, makes the equation a true number sentence.

EUREKA
MATH

> ### Lesson Summary
>
> In many word problems, an equation is often formed by setting an expression equal to a number. To build the expression, it is helpful to consider a few numerical calculations with just numbers first. For example, if a pound of apples costs $2, then three pounds cost $6 (2 × 3), four pounds cost $8 (2 × 4), and n pounds cost $2n$ dollars. If we had $15 to spend on apples and wanted to know how many pounds we could buy, we can use the expression $2n$ to write an equation, $2n = 15$, which can then be used to find the answer: $7\frac{1}{2}$ pounds.
>
> To determine if a number is a solution to an equation, substitute the number into the equation for the variable (letter) and check to see if the resulting number sentence is true. If it is true, then the number is a solution to the equation. For example, $7\frac{1}{2}$ is a solution to $2n = 15$ because $2\left(7\frac{1}{2}\right) = 15$.

Lesson 7:   Understanding Equations

Name _____   Date _____

1.   Check whether the given value of x is a solution to the equation. Justify your answer.

   a. $\frac{1}{3}(x + 4) = 20$ $x = 48$

   b. $3x - 1 = 5x + 10$ $x = -5\frac{1}{2}$

2.   The total cost of four pens and seven mechanical pencils is \$13.25. The cost of each pencil is 75 cents.

   a.   Using an arithmetic approach, find the cost of a pen.

b. Let the cost of a pen be p dollars. Write an expression for the total cost of four pens and seven mechanical pencils in terms of p.

c. Write an equation that could be used to find the cost of a pen.

d. Determine a value for p for which the equation you wrote in part (c) is true.

e. Determine a value for p for which the equation you wrote in part (c) is false.

© 2019 Great Minds®. eureka-math.org

1. Check whether the given value of h is a solution to the equation. Justify your answer.

$$4(2h - 3) = 6 + 2h \qquad h = 3$$

> I need to replace all of the h's with 3 and evaluate each side of the equation.

> Because both expressions are equal to 12 when $h = 3$, I know that $h = 3$ is a solution to the equation. If the value of each expression were not equal, I would know that the number substituted in for h was not a solution.

$$4(2(3) - 3) = 6 + 2(3)$$
$$4(6 - 3) = 6 + 6$$
$$4(3) = 12$$
$$12 = 12$$

Felix is trying to create a number puzzle for his friend to solve. He challenges his friend to find the mystery number. "When 8 is added to one-third of a number, the result is −2." The equation to represent the mystery number is $\frac{1}{3}x + 8 = -2$. Felix's friend tries to guess the mystery number. Her first guess is −18. Is she correct? Why or why not?

$$\frac{1}{3}x + 8 = -2$$

$$\frac{1}{3}(-18) + 8 = -2$$

$$\frac{1}{3}\left(-\frac{18}{1}\right) + 8 = -2$$

$$-\frac{18}{3} + 8 = -2$$

$$-6 + 8 = -2$$

$$2 = -2$$

> If I cannot remember how to work with integers, I can go back to the beginning of Module 2 for help.

> I know that 2 and −2 are opposites, which means they are not equal. So, this is not a true statement.

False

She is not correct. The number − 18 will not make a true statement. Therefore, it cannot be a solution.

2. The sum of three consecutive integers is 57.

   a. Find the smallest integer using a tape diagram.

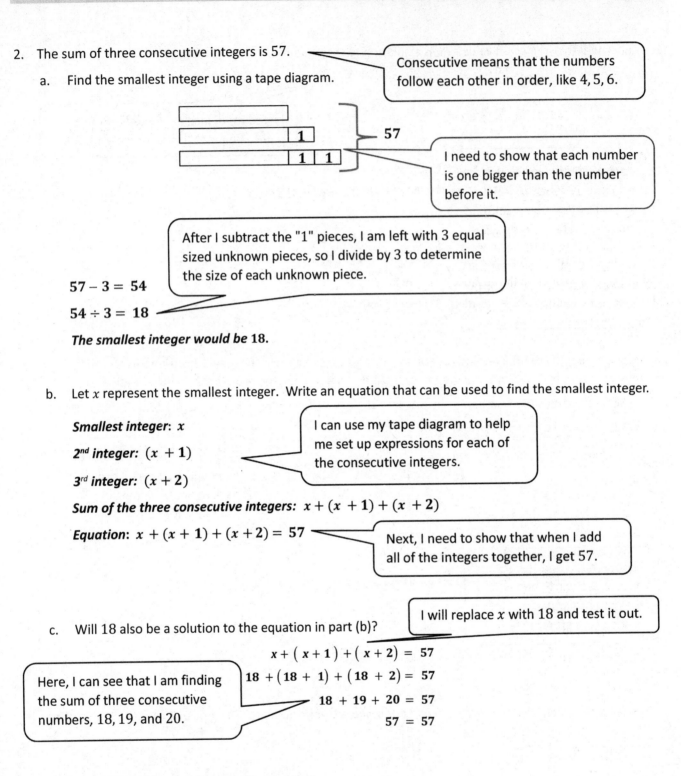

> Consecutive means that the numbers follow each other in order, like 4, 5, 6.

> I need to show that each number is one bigger than the number before it.

> After I subtract the "1" pieces, I am left with 3 equal sized unknown pieces, so I divide by 3 to determine the size of each unknown piece.

$57 - 3 = 54$

$54 \div 3 = 18$

The smallest integer would be 18.

   b. Let x represent the smallest integer. Write an equation that can be used to find the smallest integer.

   Smallest integer: x

   2nd integer: $(x + 1)$

   > I can use my tape diagram to help me set up expressions for each of the consecutive integers.

   3rd integer: $(x + 2)$

   Sum of the three consecutive integers: $x + (x + 1) + (x + 2)$

   Equation: $x + (x + 1) + (x + 2) = 57$

   > Next, I need to show that when I add all of the integers together, I get 57.

   c. Will 18 also be a solution to the equation in part (b)?

   > I will replace x with 18 and test it out.

   $$x + (x + 1) + (x + 2) = 57$$
   $$18 + (18 + 1) + (18 + 2) = 57$$
   $$18 + 19 + 20 = 57$$
   $$57 = 57$$

   > Here, I can see that I am finding the sum of three consecutive numbers, 18, 19, and 20.

   Yes, 18 is also a solution to the equation.

1. Check whether the given value is a solution to the equation.

   a. $4n - 3 = -2n + 9$ $n = 2$

   b. $9m - 19 = 3m + 1$ $m = \dfrac{10}{3}$

   c. $3(y + 8) = 2y - 6$ $y = 30$

2. Tell whether each number is a solution to the problem modeled by the following equation.

   Mystery Number: Five more than -8 times a number is 29. What is the number?

   Let the mystery number be represented by n.
   The equation is $5 + (-8)n = 29$.

   a. Is 3 a solution to the equation? Why or why not?

   b. Is -4 a solution to the equation? Why or why not?

   c. Is -3 a solution to the equation? Why or why not?

   d. What is the mystery number?

3. The sum of three consecutive integers is 36.

   a. Find the smallest integer using a tape diagram.

   b. Let n represent the smallest integer. Write an equation that can be used to find the smallest integer.

   c. Determine if each value of n below is a solution to the equation in part (b).

   $n = 12.5$

   $n = 12$

   $n = 11$

4. Andrew is trying to create a number puzzle for his younger sister to solve. He challenges his sister to find the mystery number. "When 4 is subtracted from half of a number the result is 5." The equation to represent the mystery number is $\dfrac{1}{2}m - 4 = 5$. Andrew's sister tries to guess the mystery number.

   a. Her first guess is 30. Is she correct? Why or why not?

   b. Her second guess is 2. Is she correct? Why or why not?

   c. Her final guess is $4\dfrac{1}{2}$. Is she correct? Why or why not?

Lesson 7:   Understanding Equations

Opening Exercise

Recall and summarize the if-then moves.

Write $3 + 5 = 8$ in as many true equations as you can using the if-then moves. Identify which if-then move you used.

Example 1

Julia, Keller, and Israel are volunteer firefighters. On Saturday, the volunteer fire department held its annual coin drop fundraiser at a streetlight. After one hour, Keller had collected $42.50 more than Julia, and Israel had collected $15 less than Keller. The three firefighters collected $125.95 in total. How much did each person collect?

Find the solution using a tape diagram.

Julia [x]

keller [x][42.50]

Israel [x][27.15]   } $125.95

$x + x + 42.50 + x$
27.50

$3x + 70 = 125.95$
$-70 \quad -70.00$
$\overline{3x + 0^{0} = 55.95}$ over 3

$x = 18.65$ 3

Julia
Keller 18.65 + 42.50 = 61.15
Israel 18.65 + 27.50 = 46.15

What were the operations we used to get our answer?

add/sus/divide

The amount of money Julia collected is j dollars. Write an expression to represent the amount of money Keller collected in dollars.

$42.50 + j$

Using the expressions for Julia and Keller, write an expression to represent the amount of money Israel collected in dollars.

$= 5 + j$

$j + (42.5 - 15)$

Using the expressions written above, write an equation in terms of j that can be used to find the amount each person collected.

$j + (j + 42.50) + j + (j + 27.50) = 125.95$

         K   I   I

Solve the equation written above to determine the amount of money each person collected and describe any if-then moves used.

$j + (j + 42.50) + j + (j + 27.50) = 125.95$

$3j + 70 = 125.95$

$\div 3 \quad -70 \quad -70.00$

subtract to find keller because keller is least

$55.95 \div 3 = 18.65$

$j = \$ 18.65$

Example 2

You are designing a rectangular pet pen for your new baby puppy. You have 30 feet of fence barrier. You decide that you would like the length to be $6\frac{1}{3}$ feet longer than the width.

Draw and label a diagram to represent the pet pen. Write expressions to represent the width and length of the pet pen.

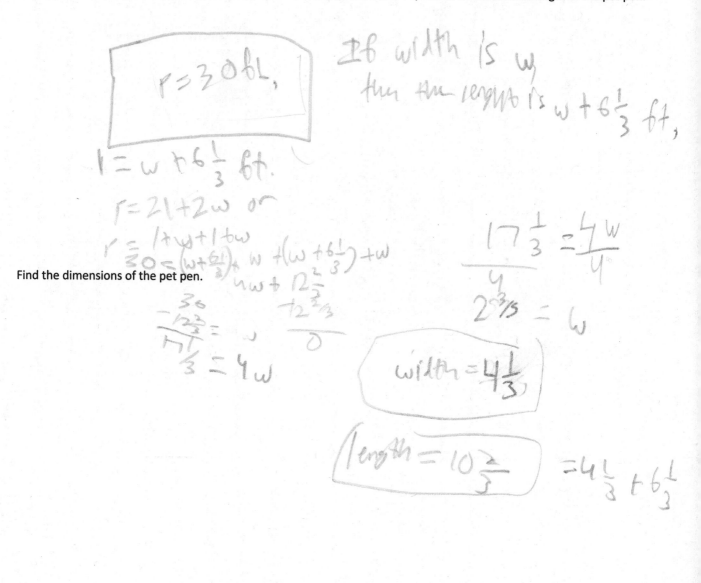

Find the dimensions of the pet pen.

Example 3

Nancy's morning routine involves getting dressed, eating breakfast, making her bed, and driving to work. Nancy spends $\frac{1}{3}$ of the total time in the morning getting dressed, 10 minutes eating breakfast, 5 minutes making her bed and the remaining time driving to work. If Nancy spent $35\frac{1}{2}$ minutes getting dressed, eating breakfast, and making her bed, how long was her drive to work?

Write and solve this problem using an equation. Identify the if-then moves used when solving the equation.

P
E
M
D
A
S

$$\text{If total time is } 35\tfrac{1}{2}$$
$$\text{then dress time is } 35\tfrac{1}{2} \times \tfrac{1}{3} \text{ or } 35.\cancel{8} \div 3$$

$$\tfrac{1}{3}x + 10 + 5 = 35\tfrac{1}{2}$$

$$\tfrac{1}{3}x + 15 = 35\tfrac{1}{2}$$
$$\phantom{\tfrac{1}{3}x +}\; -15 \quad -15$$

$$\tfrac{3}{1} \cdot \tfrac{1}{3}x = 20\tfrac{1}{2}\left(\tfrac{3}{1}\right)$$

$$x = 61\tfrac{1}{2} \quad (\text{total routine of all steps})$$

$$61.5$$
$$-35.5$$
$$\overline{26.0}$$

$$\boxed{26 \text{ mins}}$$

Is your answer reasonable? Explain.

Less than total
time so Yes

Example 4

The total number of participants who went on the seventh-grade field trip to the Natural Science Museum consisted of all of the seventh-grade students and 7 adult chaperones. Two-thirds of the total participants rode a large bus, and the rest rode a smaller bus. If 54 people rode the large bus, how many students went on the field trip?

$$s + 7 = \text{total participants}$$

$$\left(\frac{3}{2}\right)\frac{2}{3}(s+7) = 54\left(\frac{3}{2}\right)(\text{large bus passengers})$$

$$\begin{array}{r} s + 7 = 81 \\ -7 \quad -7 \\ \hline 74 \end{array}$$

$$s = 74 \text{ students}$$

Lesson Summary

Algebraic Approach: To *solve an equation* algebraically means to use the properties of operations and if-then moves to simplify the equation into a form where the solution is easily recognizable. For the equations we are studying this year (called linear equations), that form is an equation that looks like $x = a \; number$, where the number is the solution.

If-Then Moves: If x is a solution to an equation, it will continue to be a solution to the new equation formed by adding or subtracting a number from both sides of the equation. It will also continue to be a solution when both sides of the equation are multiplied by or divided by a nonzero number. We use these if-then moves to make zeros and ones in ways that simplify the original equation.

Useful First Step: If one is faced with the task of finding a solution to an equation, a useful first step is to collect like terms on each side of the equation.

Name _____   Date _____

Mrs. Canale's class is selling frozen pizzas to earn money for a field trip. For every pizza sold, the class makes $5.35. They have already earned $182.90 toward their $750 goal. How many more pizzas must they sell to earn $750? Solve this problem first by using an arithmetic approach, then by using an algebraic approach. Compare the calculations you made using each approach.

1. Four times the sum of three consecutive odd integers is -84. Find the integers.

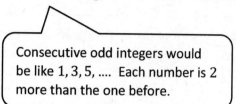

Consecutive odd integers would be like $1, 3, 5, \ldots$. Each number is 2 more than the one before.

Let n represent the first odd integer; then $n + 2$ and $n + 4$ represent the next two consecutive odd integers.

$$4\big(n + (n + 2) + (n + 4)\big) = -84$$
$$4(3n + 6) = -84$$
$$12n + 24 = -84$$
$$12n + 24 - 24 = -84 - 24$$
$$12n = -108$$
$$n = -9$$

I need to collect like terms and use the distributive property when solving.

I can go back to the expressions for each integer and substitute in the -9 for the value of n to determine the other consecutive integers.

$$-9 + 2 = -7$$
$$-9 + 4 = -5$$

The integers are $-9, -7,$ and -5.

2. A number is 11 greater than $\frac{2}{3}$ another number. If the sum of the numbers is 31, find the numbers.

> I only want to use one variable, so I need to write an expression for the first number based on how it is related to the other number.

Let n represent a number; then $\frac{2}{3}n + 11$ represents the other number.

> Rewriting some of the terms in equivalent forms, like n as $\frac{3}{3}n$, will make it easier to collect like terms.

$$n + \left(\frac{2}{3}n + 11\right) = 31$$

$$\left(n + \frac{2}{3}n\right) + 11 = 31$$

$$\left(\frac{3}{3}n + \frac{2}{3}n\right) + 11 = 31$$

$$\frac{5}{3}n + 11 = 31$$

$$\frac{5}{3}n + 11 - 11 = 31 - 11$$

$$\frac{5}{3}n + 0 = 20$$

$$\frac{5}{3}n = 20$$

$$\frac{3}{5} \cdot \frac{5}{3}n = \frac{3}{5} \cdot 20$$

$$1n = 3 \cdot 4$$

$$n = 12$$

Since the numbers sum to 31, they are 12 and 19.

Lesson 8:   Using If-Then Moves in Solving Equations

3. Lukas filled 6.5 more boxes than Charlotte, and Xin filled 8 fewer than Lukas. Together, they filled 50 boxes. How many boxes did each person fill?

> There are three different people mentioned here. Lukas is being compared to Charlotte, but I don't know anything about how many Charlotte filled, so I will call it n.

Let n represent the number of boxes Charlotte filled.

Then, $(n + 6.5)$ will represent the number of boxes Lukas filled.

And $((n + 6.5) - 8)$ or $(n - 1.5)$ will represent the number of boxes Xin filled.

> Simplifying the expression for Xin now will make it easier to work with later.

$$n + (n + 6.5) + (n + 1.5) = 50$$
$$n + n + n + 6.5 - 1.5 = 50$$
$$3n + 5 = 50$$
$$3n + 5 - 5 = 50 - 5$$
$$3n = 45$$
$$\left(\frac{1}{3}\right)(3n) = \left(\frac{1}{3}\right)(45)$$
$$n = 15$$

> The equation I write should be Charlotte's expression + Lukas' expression + Xin's expression = 50. Then I replace the people's names with expressions that represent how many boxes they filled.

$$15 + 6.5 = 21.5$$
$$15 - 1.5 = 13.5$$
$$15 + 21.5 + 13.5 = 50$$

If the total number of boxes filled was 50, then Charlotte filled 15 boxes, Lukas filled 21.5 boxes, and Xin filled 13.5 boxes.

4. A preschool teacher plans her class to include 30 minutes on the playground, $\frac{1}{4}$ of the daily class time on a craft/project, and the remaining practice time working on skills, reading, and math. The teacher planned 75 minutes for the playground and craft/project time. How long, in hours, is a day of preschool?

The duration of the entire preschool day: x hours

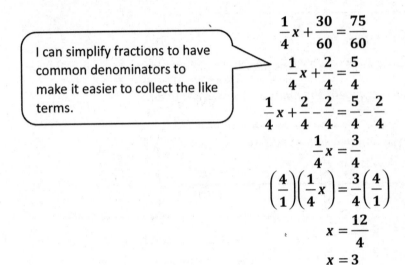

I can simplify fractions to have common denominators to make it easier to collect the like terms.

The problem says to give the time in hours. I know that there are 60 minutes in an hour. I can write the minutes given in terms of hours by placing them in a fraction with 60 in the denominator.

$$\frac{1}{4}x + \frac{30}{60} = \frac{75}{60}$$

$$\frac{1}{4}x + \frac{2}{4} = \frac{5}{4}$$

$$\frac{1}{4}x + \frac{2}{4} - \frac{2}{4} = \frac{5}{4} - \frac{2}{4}$$

$$\frac{1}{4}x = \frac{3}{4}$$

$$\left(\frac{4}{1}\right)\left(\frac{1}{4}x\right) = \frac{3}{4}\left(\frac{4}{1}\right)$$

$$x = \frac{12}{4}$$

$$x = 3$$

Preschool is 3 hours long each day.

Lesson 8:   Using If-Then Moves in Solving Equations

Write and solve an equation for each problem.

1. The perimeter of a rectangle is 30 inches. If its length is three times its width, find the dimensions.

    $30 = (3w) + w + (3w) + w + (3w)$; $30 = 6 + 5w$, $w = 3.75$

2. A cell phone company has a basic monthly plan of $40 plus $0.45 for any minutes used over 700. Before receiving his statement, John saw he was charged a total of $48.10. Write and solve an equation to determine how many minutes he must have used during the month. Write an equation without decimals. mc minutes

    ?

3. A volleyball coach plans her daily practices to include 10 minutes of stretching, $\frac{2}{3}$ of the entire practice scrimmaging, and the remaining practice time working on drills of specific skills. On Wednesday, the coach planned 100 minutes of stretching and scrimmaging. How long, in hours, is the entire practice?

    $\frac{2}{3} x + 10 = 100 \cdot \frac{3}{2} \leq x = 150$ hours

4. The sum of two consecutive even numbers is 54. Find the numbers.

    22 and 22

5. Justin has $7.50 more than Eva, and Emma has $12 less than Justin. Together, they have a total of $63.00. How much money does each person have? a = eva mon
    $a + a + 7.50 + a - 4.5 = 63.3$ $7.4 - 4.5 = 3$
    $3a + 3 = 63 - 6.3$ $a = 20$ $j = 27.50$, $e = 15.50$

6. Barry's mountain bike weighs 6 pounds more than Andy's. If their bikes weigh 42 pounds altogether, how much does Barry's bike weigh? Identify the if-then moves in your solution. a = andy
    $a + \frac{6}{6} = 42 - 3(1)$ barry = 36

7. Trevor and Marissa together have 26 T-shirts to sell. If Marissa has 6 fewer T-shirts than Trevor, find how many T-shirts Trevor has. Identify the if-then moves in your solution. $t = $ trevor
    $26 = t + (t - 6) = 26 = 2t - \frac{6}{6}$

8. A number is $\frac{1}{7}$ of another number. The difference of the numbers is 18. (Assume that you are subtracting the smaller number from the larger number.) Find the numbers.
    $18 = n - (\frac{1}{7}n)$ $\frac{2}{7}$

9. A number is 6 greater than $\frac{1}{2}$ another number. If the sum of the numbers is 21, find the numbers. n
    $21 = n + (6 + \frac{1}{2}n) = 21 = \frac{3}{2}n + 6 = n = 5$

10. Kevin is currently twice as old now as his brother. If Kevin was 8 years old 2 years ago, how old is Kevin's brother now?

    ?

11. The sum of two consecutive odd numbers is 156. What are the numbers?

    79 + 77

12. If n represents an odd integer, write expressions in terms of n that represent the next three consecutive odd integers. If the four consecutive odd integers have a sum of 56, find the numbers.

    $4n = 56$

13. The cost of admission to a history museum is $3.25 per person over the age of 3; kids 3 and under get in for free. If the total cost of admission for the Warrick family, including their two 6-month old twins, is $19.50, find how many family members are over 3 years old.

    $6 + a$ $19.50 \div 3.25$

EUREKA
MATH

Lesson 8:   Using If-Then Moves in Solving Equations   111

© 2019 Great Minds®. eureka-math.org

14. Six times the sum of three consecutive odd integers is -18. Find the integers.

15. I am thinking of a number. If you multiply my number by 4, add -4 to the product, and then take $\frac{1}{3}$ of the sum, the result is -6. Find my number.

16. A vending machine has twice as many quarters in it as dollar bills. If the quarters and dollar bills have a combined value of $96.00, how many quarters are in the machine?

Lesson 8:   Using If-Then Moves in Solving Equations

EUREKA
MATH

Opening Exercise

Heather practices soccer and piano. Each day she practices piano for 2 hours. After 5 days, she practiced both piano and soccer for a total of 20 hours. Assuming that she practiced soccer the same amount of time each day, how many hours per day, h, did Heather practice soccer?

$h =$ hours playing soccer

$$5(h+2)$$

$= 2 \text{ hours}$

$$5h + 10 = 20$$
$$\frac{-10}{5} \quad \frac{-10}{10}$$
$$5 \div 5h = 10 \div 5$$
$$h = 2$$

Over 5 days, Jake practices piano for a total of 2 hours. Jake practices soccer for the same amount of time each day. If he practiced piano and soccer for a total of 20 hours, how many hours, h, per day did Jake practice soccer?

$h =$ soccer hours

$$5h + 2 = 20$$
$$\frac{-2}{0} \quad \frac{-2}{18}$$
$$5 \div 5h = \frac{18}{18} \div 5$$
$$h = 3\frac{3}{5} \text{ or}$$
$$h = 3.6$$

Example 1

Fred and Sam are a team in the local 138.2 mile bike-run-athon. Fred will compete in the bike race, and Sam will compete in the run. Fred bikes at an average speed of 8 miles per hour and Sam runs at an average speed of 4 miles per hour. The bike race begins at 6:00 a.m., followed by the run. Sam predicts he will finish the run at 2:33 a.m. the next morning.

a. How many hours will it take them to complete the entire bike-run-athon?

$$20:33 \text{ or } 20.55 \text{ hrs}$$

b. If t is how long it takes Fred to complete the bike race, in hours, write an expression to find Fred's total distance.

$$d = r \cdot t$$
$$d = 8t$$

c. Write an expression, in terms of t to express Sam's time.

$$20.55 - t$$

d. Write an expression, in terms of t, that represents Sam's total distance.

$$d = 4(20.55 - t)$$
$$d = 82.20 - 4t$$

e. Write and solve an equation using the total distance both Fred and Sam will travel.

$$(8t) + (82.20 - 4t) = 138.2$$
$$\underset{13}{(4t)} + 82.2 = 138.2$$
$$\begin{array}{rr} -82.2 & -82.2 \\ \hline 0 & 56.0 \end{array}$$

$$4\,|\,4t = 56$$

Fred $t = 14$, Sam $20.55 - 14 = 6.55$

EUREKA MATH

f. How far will Fred bike, and how much time will it take him to complete his leg of the race?

$8(14) = 112 \text{ mile}$

g. How far will Sam run, and how much time will it take him to complete his leg of the race?

$t = 4(20.55 - (t))$

$t = 14$

$4 \cdot 6.55 = \boxed{26.2 \text{ mile}}$

Total Time (hours)	Fred's Time (hours)	Sam's Time (hours)
10	6	4
15	12	3
20	8	12
18.35	8	10.35
20.55	t	$20.55 - t$

X	rate	time	distance
Fred	8	t	$8t$
Sam	4	$20.55 - t$	$4(20.55 - t)$

Example 2

Shelby is seven times as old as Bonnie. If in 5 years, the sum of Bonnie and Shelby's ages is 98, find Bonnie's present age. Use an algebraic approach.

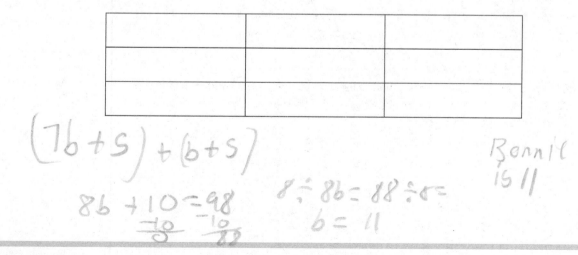

$(7b + 5) + (b + 5)$

$8b + 10 = 98$
$\quad -10 \quad -10$

$8 \div 8b = 88 \div 8 =$
$b = 11$

Bonnie is 11

Group 1: Where can you buy a ruler that is 3 feet long?

_____	_____	_____	_____		_____	_____	_____	_____
3	$4\frac{1}{2}$	3.5	-1		-2	19	18.95	4.22

What value(s) of z makes the equation $\dfrac{7}{6}z + \dfrac{1}{3} = -\dfrac{5}{6}$ true; $z = -1$, $z = 2$, $z = 1$, or $z = -\dfrac{36}{63}$?	D
Find the smaller of 2 consecutive integers if the sum of the smaller and twice the larger is -4.	S
Twice the sum of a number and -6 is -6. Find the number.	Y
Logan is 2 years older than Lindsey. Five years ago, the sum of their ages was 30. Find Lindsey's current age.	A
The total charge for a taxi ride in NYC includes an initial fee of $3.75 plus $1.25 for every $\dfrac{1}{2}$ mile traveled. Jodi took a taxi, and the ride cost her exactly $12.50. How many <u>miles</u> did she travel in the taxi?	R
The perimeter of a triangular garden with 3 equal sides is 12.66 feet. What is the length of each side of the garden?	E
A car travelling at 60 mph leaves Ithaca and travels west. Two hours later, a truck travelling at 55 mph leaves Elmira and travels east. Altogether, the car and truck travel 407.5 miles. How many hours does the car travel?	A
The Cozo family has 5 children. While on vacation, they went to a play. They bought 5 tickets at the child's price of $10.25 and 2 tickets at the adult's price. If they spent a total of $89.15, how much was the price of each adult ticket?	L

EUREKA
MATH

Group 2: Where do fish keep their money?

___	___	___	___	___		___	___	___	___
2	−1	10	8	2		−6	5	50	$\frac{1}{8}$

What value of z makes the equation $\frac{2}{3}z - \frac{1}{2} = -\frac{5}{12}$ true; $z = -1$, $z = 2$, $z = \frac{1}{8}$, or $z = -\frac{1}{8}$?	K
Find the smaller of 2 consecutive even integers if the sum of twice the smaller integer and the larger integers is −16.	B
Twice the difference of a number and −3 is 4. Find the number.	I
Brooke is 3 years younger than Samantha. In five years, the sum of their ages will be 29. Find Brooke's age.	E
Which of the following equations is equivalent to $4.12x + 5.2 = 8.23$? (1) $412x + 52 = 823$ (2) $412x + 520 = 823$ (3) $9.32x = 8.23$ (4) $0.412x + 0.52 = 8.23$	R
The length of a rectangle is twice the width. If the perimeter of the rectangle is 30 units, find the area of the garden.	N
A car traveling at 70 miles per hour traveled one hour longer than a truck traveling at 60 miles per hour. If the car and truck traveled a total of 330 miles, for how many hours did the car and truck travel altogether?	A
Jeff sold half of his baseball cards then bought sixteen more. He now has 21 baseball cards. How many cards did he begin with?	V

Group 3: The more you take, the more you leave behind. What are they?

___	___	___	___	___	___	___	___	___
8	11.93	368	$1\frac{5}{6}$	10.50	$2\frac{1}{2}$	$3\frac{5}{6}$	21	4

An apple has 80 calories. This is 12 less than $\frac{1}{4}$ the number of calories in a package of candy. How many calories are in the candy?	O
The ages of 3 brothers are represented by consecutive integers. If the oldest brother's age is decreased by twice the youngest brother's age, the result is −19. How old is the youngest brother?	P
A carpenter uses 3 hinges on every door he hangs. He hangs 4 doors on the first floor and x doors on the second floor. If he uses 36 hinges total, how many doors did he hang on the second floor?	F
Kate has $12\frac{1}{2}$ pounds of chocolate. She gives each of her 5 friends x pounds each and has $3\frac{1}{3}$ pounds left over. How much did she give each of her friends?	T
A room is 20 feet long. If a couch that is $12\frac{1}{3}$ feet long is to be centered in the room, how big of a table can be placed on either side of the couch?	E
Which equation is equivalent to $\frac{1}{4}x + \frac{1}{5} = 2$? (1) $4x + 5 = \frac{1}{2}$ (2) $\frac{2}{9}x = 2$ (3) $5x + 4 = 18$ (4) $5x + 4 = 40$	S
During a recent sale, the first movie purchased cost $29, and each additional movie purchased costs m dollars. If Jose buys 4 movies and spends a total of $64.80, how much did each additional movie cost?	O
The Hipster Dance company purchases 5 bus tickets that cost $150 each, and they have 7 bags that cost b dollars each. If the total bill is $823.50, how much does each bag cost?	S
The weekend before final exams, Travis studied 1.5 hours for his science exam, $2\frac{1}{4}$ hours for his math exam, and h hours each for Spanish, English, and social studies. If he spent a total of $11\frac{1}{4}$ hours studying, how much time did he spend studying for Spanish?	T

Name _____   Date _____

1. Brand A scooter has a top speed that goes 2 miles per hour faster than Brand B. If after 3 hours, Brand A scooter traveled 24 miles at its top speed, at what rate did Brand B scooter travel at its top speed if it traveled the same distance? Write an equation to determine the solution. Identify the if-then moves used in your solution.

2. At each scooter's top speed, Brand A scooter goes 2 miles per hour faster than Brand B. If after traveling at its top speed for 3 hours, Brand A scooter traveled 40.2 miles, at what rate did Brand B scooter travel if it traveled the same distance as Brand A? Write an equation to determine the solution and then write an equivalent equation using only integers.

1. Holly's grandfather is 52 years older than her. In 7 years, the sum of their ages will be 70. Find Holly's present age.

    Let x represents Holly's age now in years.

	Now	*7 years later*
Holly	x	$x + 7$
Grandfather	$x + 52$	$(x + 52) + 7$

The question mentions now and 7 years later. I can make a table to organize the information provided to help me create an equation to model the situation.

$$x + 7 + x + 52 + 7 = 70$$
$$x + x + 7 + 52 + 7 = 70$$
$$2x + 66 = 70$$
$$2x + 66 - 66 = 70 - 66$$
$$2x = 4$$
$$\left(\frac{1}{2}\right)(2x) = \left(\frac{1}{2}\right)(4)$$
$$x = 2$$

I was given the sum after 7 years, so I will use the third column in the table to form the equation.

Holly's present age is 2 years old.

2. The sum of two numbers is 63, and their difference is 7. Find the numbers.

Let x represent one of the two numbers.

Let 63 – x represent the other number.

This question tells me about the sum and the difference. I use the sum to set up the "let" statements and the difference to write the equation that models the situation.

If two numbers have a sum of 63, and I take one number away from 63, I will get the other number. I can check this by adding them together.

$$x + (63 - x) = x - x + 63 = 63$$

$$x - (63 - x) = 7$$
$$x + (-(63 - x)) = 7$$
$$x + (-63 + x) = 7$$
$$x + (-63) + x = 7$$
$$2x - 63 = 7$$
$$2x - 63 + 63 = 7 + 63$$
$$2x = 70$$
$$\left(\frac{1}{2}\right)(2x) = \left(\frac{1}{2}\right)(70)$$
$$x = 35$$

$$63 - 35 = 28$$

The numbers are 35 and 28.

3. Carmen is planning a party to introduce people to her new products for sale. She bought 500 gifts bags to hold party favors and 500 business cards. Each gift bag costs 57 cents more than each business card. If Carmen's total order costs $315, find the cost of each gift bag and business card.

> This question gives money in cents and money in dollars, but I need common units, so I write both of the amounts in dollars.

Let b represent the cost of a business card.

Then, the cost of a gift bag in dollars is $b + 0.57$.

> Because she bought 500 of both items, I can use the distributive property to write this equation.

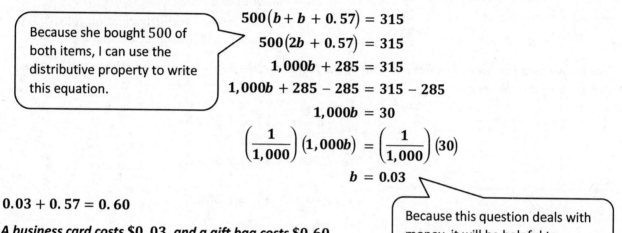

$$500(b + b + 0.57) = 315$$
$$500(2b + 0.57) = 315$$
$$1,000b + 285 = 315$$
$$1,000b + 285 - 285 = 315 - 285$$
$$1,000b = 30$$
$$\left(\frac{1}{1,000}\right)(1,000b) = \left(\frac{1}{1,000}\right)(30)$$
$$b = 0.03$$

> Because this question deals with money, it will be helpful to convert from fraction to decimal.

$$0.03 + 0.57 = 0.60$$

A business card costs $0.03, and a gift bag costs $0.60.

4. A group of friends left for vacation in two vehicles at the same time. One car traveled an average speed of 4 miles per hour faster than the other. When the first car arrived at the destination after $8\frac{1}{4}$ hours of driving, both cars had driven a total of 1,006.5 miles. If the second car continues at the same average speed, how much time, to the nearest minute, will it take before the second car arrives?

> The 1,006.5 miles doesn't represent the total miles driven for the whole trip. Instead, this is the amount that both cars drove in $8\frac{1}{4}$ hours. The second car hasn't arrived at the destination yet.

Let r represent the speed in miles per hour of the faster car; then r − 4 represents the speed in miles per hour of the slower car.

$$8\frac{1}{4}(r) + 8\frac{1}{4}(r-4) = 1{,}006.5$$
$$8\frac{1}{4}(r+r-4) = 1{,}006.5$$
$$8\frac{1}{4}(2r-4) = 1{,}006.5$$
$$\frac{33}{4}(2r-4) = 1{,}006.5$$
$$\frac{4}{33}\cdot\frac{33}{4}(2r-4) = \frac{4}{33}\cdot 1{,}006.5$$
$$2r-4 = 122$$
$$2r-4+4 = 122+4$$
$$2r = 126$$
$$\left(\frac{1}{2}\right)(2r) = \left(\frac{1}{2}\right)(126)$$
$$r = 63$$

> I can use the formula
> distance = rate × time to help me set up an equation for this problem. This will first help me to determine how fast each vehicle was going.

> I can think of 1,006.5 as $\frac{1{,}006.5}{1}$ so that I can multiply the two factors.

The average speed of the faster car is 63 miles per hour, so the average speed of the slower car is 59 miles per hour.

$$d = 59\cdot 8\frac{1}{4}$$
$$d = 59\cdot\frac{33}{4}$$
$$d = 486.75$$

Lesson 9: Using If-Then Moves in Solving Equations

EUREKA MATH

The slower car traveled 486.75 miles in $8\frac{1}{4}$ hours.

$1,006.5 - 486.75 = 519.75$

The faster car traveled 519.75 miles in $8\frac{1}{4}$ hours.

The slower car traveled 486.75 miles in $8\frac{1}{4}$ hours.

The remainder of the slower car's trip is 33 miles because $519.75 - 486.75 = 33$.

Now that I know the rate and distance the second car still needs to travel, I can use $d = rt$ again to solve for the time.

$$33 = 59\,(t)$$
$$\frac{1}{59}(33) = \frac{1}{59}(59)(t)$$
$$\frac{33}{59} = t$$

This time is in hours. To convert to minutes, multiply by 60 because there are 60 minutes in an hour.

$$\frac{33}{59} \cdot 60 = \frac{1980}{59} \approx 34$$

The slower car will arrive approximately 34 minutes after the faster car.

Lesson 9:   Using If-Then Moves in Solving Equations

5. Lucien bought a certain brand of fertilizer for his garden at a unit price of $1.25 per pound. The total cost of the fertilizer left him with $5. He wanted to buy the same weight of a better brand of fertilizer, but at $2.10 per pound, he would have been $80 short of the total amount due. How much money did Lucien have to buy fertilizer?

> From the word problem, I can determine the difference in how much money is left between buying the cheaper or the more expensive product.
> $$5 - (-80) = 5 + 80 = 85.$$

The difference in the costs is $\$85.00$ *for the same weight in fertilizer.*

Let w represent the weight in pounds of fertilizer.

$$2.10w - 1.25w = 85$$
$$0.85w = 85$$
$$\frac{85}{100}w = 85$$
$$\frac{100}{85} \cdot \frac{85}{100}w = 85 \cdot \frac{100}{85}$$
$$1w = 100$$
$$w = 100$$

> I can use the difference in the price per pound with the difference in the amount of money Lucien will have to help me determine how much was bought.

Lucien bought 100 *pounds of fertilizer.*

$$\text{Cost} = \text{unit price} \cdot \text{weight}$$
$$\text{Cost} = \big(\$1.25 \text{ per pound}\big) \cdot \big(100 \text{ pounds}\big)$$
$$\text{Cost} = \$125.00$$

Lucien paid $\$125$ *for* 100 *pounds of fertilizer. Lucien had* $\$5$ *left after his purchase, so he started with* $\$125 + \$5 = \$130.$

> If he would have had $5 left after paying, I need to add that to the $125 he paid for the fertilizer to determine how much he started with.

1. A company buys a digital scanner for $12,000. The value of the scanner is $12,000\left(1 - \frac{n}{5}\right)$ after n years. The company has budgeted to replace the scanner when the trade-in value is $2,400. After how many years should the company plan to replace the machine in order to receive this trade-in value?

2. Michael is 17 years older than John. In 4 years, the sum of their ages will be 49. Find Michael's present age.

   27-27=14 / J=12, p+1=29 Michael age is 29

3. Brady rode his bike 70 miles in 4 hours. He rode at an average speed of 17 mph for t hours and at an average rate of speed of 22 mph for the rest of the time. How long did Brady ride at the slower speed? Use the variable t to represent the time, in hours, Brady rode at 17 mph.

4. Caitlan went to the store to buy school clothes. She had a store credit from a previous return in the amount of $39.58. If she bought 4 of the same style shirt in different colors and spent a total of $52.22 after the store credit was taken off her total, what was the price of each shirt she bought? Write and solve an equation with integer coefficients.

   x = price of 1 shirt
   $4x + 39.58 = 52.22$ $4x \div 4, = x = 3.16$

5. A young boy is growing at a rate of 3.5 cm per month. He is currently 90 cm tall. At that rate, in how many months will the boy grow to a height of 132 cm?

6. The sum of a number, $\frac{1}{6}$ of that number, $2\frac{1}{2}$ of that number, and 7 is $12\frac{1}{2}$. Find the number.

7. The sum of two numbers is 33 and their difference is 2. Find the numbers.

8. Aiden refills three token machines in an arcade. He puts twice the number of tokens in machine A as in machine B, and in machine C, he puts $\frac{3}{4}$ what he put in machine A. The three machines took a total of 18,324 tokens. How many did each machine take?

9. Paulie ordered 250 pens and 250 pencils to sell for a theatre club fundraiser. The pens cost 11 cents more than the pencils. If Paulie's total order costs $42.50, find the cost of each pen and pencil.

10. A family left their house in two cars at the same time. One car traveled an average of 7 miles per hour faster than the other. When the first car arrived at the destination after $5\frac{1}{2}$ hours of driving, both cars had driven a total of 599.5 miles. If the second car continues at the same average speed, how much time, to the nearest minute, will it take before the second car arrives?

11. Emily counts the triangles and parallelograms in an art piece and determines that altogether, there are 42 triangles and parallelograms. If there are 150 total sides, how many triangles and parallelograms are there?

12. Stefan is three years younger than his sister Katie. The sum of Stefan's age 3 years ago and $\frac{2}{3}$ of Katie's age at that time is 12. How old is Katie now?

13. Lucas bought a certain weight of oats for his horse at a unit price of $0.20 per pound. The total cost of the oats left him with $1. He wanted to buy the same weight of enriched oats instead, but at $0.30 per pound, he would have been $2 short of the total amount due. How much money did Lucas have to buy oats?

Angle Facts and Definitions

Name of Angle Relationship	Angle Fact	Diagram
Adjacent Angles	Two angels next to each other with common common side tc.	
Vertical Angles (vert. ∠s)	r side form two pairs of opposite rys.	
Angles on a Line (∠s on a line)	Sum of 2 angles= 180°	
Angles at a Point (∠s at a point)	mesoure of angle with 3 arms = to 360 $a + b + c = 360$	

Opening Exercise

Use the diagram to complete the chart.

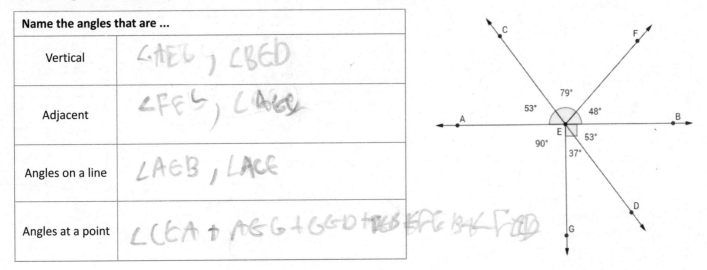

Name the angles that are ...	
Vertical	∠AEC , ∠BED
Adjacent	∠FEB , ∠AEC
Angles on a line	∠AEB , ∠ACE
Angles at a point	∠CEA + AEG + GED +DEB EFG BEF FEB

Example 1

Estimate the measurement of x. _____ 48 _____

In a complete sentence, describe the angle relationship in the diagram.

Angels on a line

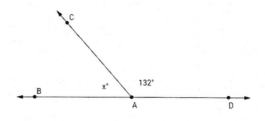

Write an equation for the angle relationship shown in the figure and solve for x. Then, find the measures of ∠BAC and confirm your answers by measuring the angle with a protractor.

Exercise 1

In a complete sentence, describe the angle relationship in the diagram.

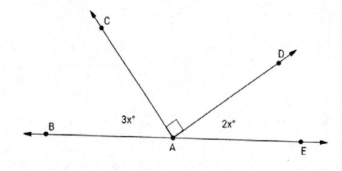

Angles at a point

Find the measurements of ∠BAC and ∠DAE.

∠BAC = 108
∠ADE = 72

Example 2

In a complete sentence, describe the angle relationship in the diagram.

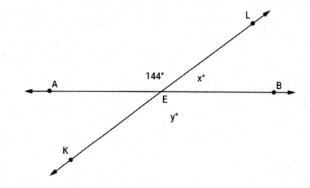

Angles on a line

Write an equation for the angle relationship shown in the figure and solve for x and y. Find the measurements of ∠LEB and ∠KEB.

y = 144

x + 144 = 180
     -144   -144

x = 36

Exercise 2

In a complete sentence, describe the angle relationships in the diagram.

Adjacent angles

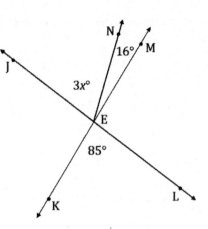

Write an equation for the angle relationship shown in the figure and solve for x.

Example 3

In a complete sentence, describe the angle relationships in the diagram.

$\angle EKG = 90$ 135 + 40 +

$\angle GKF = 135$

          Angles at
              a point

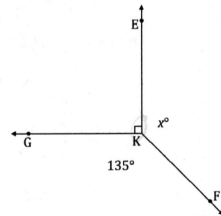

Write an equation for the angle relationship shown in the figure and solve for
x. Find the measurement of $\angle EKF$ and confirm your answers by measuring
the angle with a protractor.

$90 + 135 + x = 360$

$x + 225 = 360$

$\underline{-225 \quad -225}$

$x = 135$

$EKF = 135°$

Exercise 3

In a complete sentence, describe the angle relationships in the diagram.

Angle at a point

Find the measurement of ∠GAH.

$$329 + x = 360$$
$$-329 \qquad \frac{-329}{31}$$
$$x = 31$$

∠GAH = 32°

$$3t + 1 = 32$$

Diagram labels: F, E, 103°, 59°, G, A, (x+1)°, 167°, H

Example 4

The following two lines intersect. The ratio of the measurements of the obtuse angle to the acute angle in any adjacent angle pair in this figure is 2 : 1. In a complete sentence, describe the angle relationships in the diagram.

Angles on a line

Label the diagram with expressions that describe this relationship. Write an equation that models the angle relationship and solve for x. Find the measurements of the acute and obtuse angles.

$$2x + 1x = 180$$
$$3 \div 3x = 180 \div 3$$
$$x = 60$$
$$60 \cdot 2 = 120 - obtuse\ angle$$
$$60 \cdot 1 = 60 - acute\ angle$$

Exercise 4

The ratio of $m\angle GFH$ to $m\angle EFH$ is 2 : 3. In a complete sentence, describe the angle relationships in the diagram.

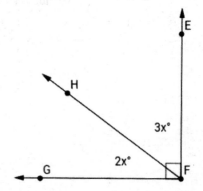

Find the measures of $\angle GFH$ and $\angle EFH$.

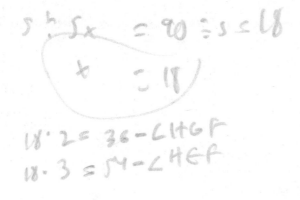

Relevant Vocabulary

ADJACENT ANGLES: Two angles $\angle BAC$ and $\angle CAD$ with a common side $\overrightarrow{AC}$ are *adjacent angles* if C belongs to the interior of $\angle BAD$.

VERTICAL ANGLES: Two angles are *vertical angles* (or *vertically opposite angles*) if their sides form two pairs of opposite rays.

ANGLES ON A LINE: The sum of the measures of adjacent *angles on a line* is 180°.

ANGLES AT A POINT: The sum of the measures of adjacent *angles at a point* is 360°.

Name _____   Date _____

In a complete sentence, describe the relevant angle relationships in the following diagram. That is, describe the angle relationships you could use to determine the value of x.

Use the angle relationships described above to write an equation to solve for x. Then, determine the measurements of $\angle JAH$ and $\angle HAG$.

For each question, use angle relationships to write an equation in order to solve for each variable. Determine the indicated angles.

1. In a complete sentence, describe the relevant angle relationships in the following diagram. Find the measurements of $\angle ABE$ and $\angle EBD$.

$\angle ABE$, $\angle EBD$, and $\angle DBC$ are angles on a line and their measures sum to $180°$.

$$3x + 5x + 28 = 180$$
$$8x + 28 = 180$$
$$8x + 28 - 28 = 180 - 28$$
$$8x = 152$$
$$\left(\frac{1}{8}\right)8x = \left(\frac{1}{8}\right)152$$
$$x = 19$$

I can see that all three of these angles form a straight line, which means the sum of these three angles must be $180°$.

$$m\angle ABE = 3(19°) = 57°$$
$$m\angle EBD = 5(19°) = 95°$$

Finding the value of x is not the answer. I need to go one step further and plug x back into the expressions and evaluate to determine the measure of each angle.

2. In a complete sentence, describe the relevant angle relationships in the following diagram. Find the measurement of $\angle WSV$.

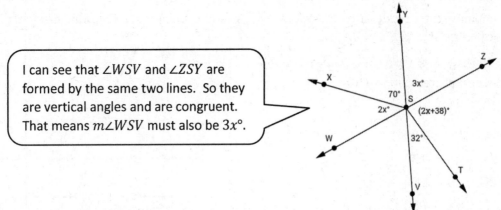

I can see that $\angle WSV$ and $\angle ZSY$ are formed by the same two lines. So they are vertical angles and are congruent. That means $m\angle WSV$ must also be $3x°$.

All of the angles in the diagram are angles at a point, and
their measures sum to 360°. $\angle ZSY$ and $\angle WSV$ are vertical angles and are of equal measurement.

$$3x + 70 + 2x + 3x + (2x + 38) + 32 = 360$$
$$3x + 2x + 3x + 2x + 70 + 38 + 32 = 360$$
$$10x + 140 = 360$$
$$10x + 140 - 140 = 360 - 140$$
$$10x = 220$$
$$\left(\frac{1}{10}\right)10x = \left(\frac{1}{10}\right)220$$
$$x = 22$$

I need to be sure to include another $3x$ term for the missing vertical angle when I set up my equation.

$$m\angle WSV = 3x° = 3(22°) = 66°$$

---

3. The ratio of the measures of three adjacent angles on a line is 1 : 4 : 7.

   a. Find the measures of the three angles.

   $m\angle 1 = x°, m\angle 2 = 4x°, m\angle 3 = 7x°$

   $x + 4x + 7x = 180$

   $\qquad\qquad 12x = 180$

   $\left(\dfrac{1}{12}\right)12x = \left(\dfrac{1}{12}\right)180$

   $\qquad\qquad\quad x = 15$

> I can use the ratio to set up an expression for each angle. I also know the measure of adjacent angles on a line must have a sum of 180°.

   $m\angle 1 = 15°$

   $m\angle 2 = 4(15°) = 60°$

   $m\angle 3 = 7(15°) = 105°$

   b. Draw a diagram to scale of these adjacent angles. Indicate the measurements of each angle.

> I can use my protractor to measure the angles accurately.

105°   60°   15°

For each question, use angle relationships to write an equation in order to solve for each variable. Determine the indicated angles. You can check your answers by measuring each angle with a protractor.

1. In a complete sentence, describe the relevant angle relationships in the following diagram. Find the measurement of ∠DAE.

25°

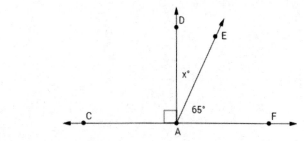

2. In a complete sentence, describe the relevant angle relationships in the following diagram. Find the measurement of ∠QPR.

13°

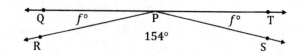

3. In a complete sentence, describe the relevant angle relationships in the following diagram. Find the measurements of ∠CQD and ∠EQF.

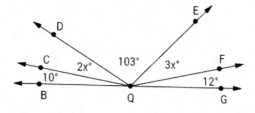

4. In a complete sentence, describe the relevant angle relationships in the following diagram. Find the measure of x.

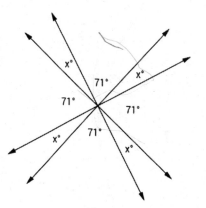

5. In a complete sentence, describe the relevant angle relationships in the following diagram. Find the measures of x and y.

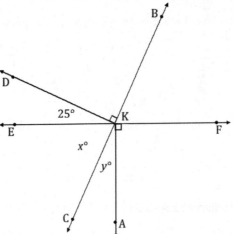

6. In a complete sentence, describe the relevant angle relationships in the following diagram. Find the measures of x and y.

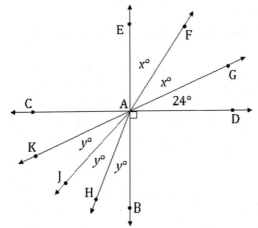

7. In a complete sentence, describe the relevant angle relationships in the following diagram. Find the measures of $\angle CAD$ and $\angle DAE$.

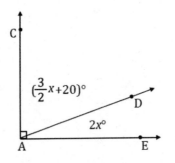

Lesson 10:   Angle Problems and Solving Equations

8. In a complete sentence, describe the relevant angle relationships in the following diagram. Find the measure of
    $\angle CQG$.

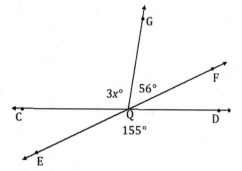

9. The ratio of the measures of a pair of adjacent angles on a line is $4 : 5$.

    a. Find the measures of the two angles.

    b. Draw a diagram to scale of these adjacent angles. Indicate the measurements of each angle.

10. The ratio of the measures of three adjacent angles on a line is $3 : 4 : 5$.

    a. Find the measures of the three angles.

    b. Draw a diagram to scale of these adjacent angles. Indicate the measurements of each angle.

Opening Exercise

a. In a complete sentence, describe the angle relationship in the diagram. Write an equation for the angle relationship shown in the figure and solve for x. Confirm your answer by measuring the angle with a protractor.

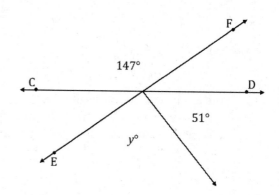

b. $\overleftrightarrow{CD}$ and $\overleftrightarrow{EF}$ are intersecting lines. In a complete sentence, describe the angle relationship in the diagram. Write an equation for the angle relationship shown in the figure and solve for y. Confirm your answer by measuring the angle with a protractor.

c. In a complete sentence, describe the angle relationship in the diagram. Write an equation for the angle relationship shown in the figure and solve for b. Confirm your answer by measuring the angle with a protractor.

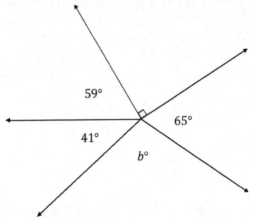

d. The following figure shows three lines intersecting at a point. In a complete sentence, describe the angle relationship in the diagram. Write an equation for the angle relationship shown in the figure and solve for z. Confirm your answer by measuring the angle with a protractor.

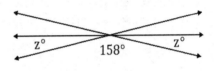

e. Write an equation for the angle relationship shown in the figure and solve for x. In a complete sentence, describe the angle relationship in the diagram. Find the measurements of $\angle EPB$ and $\angle CPA$. Confirm your answers by measuring the angles with a protractor.

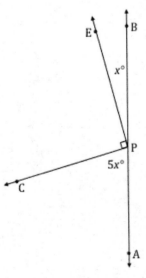

The following figure shows three lines intersecting at a point. In a complete sentence, describe the angle relationship in the diagram. Write an equation for the angle relationship shown in the figure and solve for x. Confirm your answer by measuring the angle with a protractor.

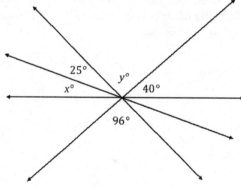

Exercise 1

The following figure shows four lines intersecting at a point. In a complete sentence, describe the angle relationships in the diagram. Write an equation for the angle relationship shown in the figure and solve for x and y. Confirm your answers by measuring the angles with a protractor.

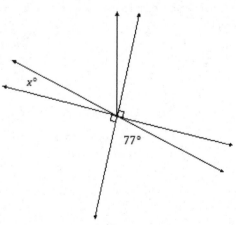

In a complete sentence, describe the angle relationships in the diagram. You may label the diagram to help describe the angle relationships. Write an equation for the angle relationship shown in the figure and solve for x. Confirm your answer by measuring the angle with a protractor.

Exercise 2

In a complete sentence, describe the angle relationships in the diagram. Write an equation for the angle relationship shown in the figure and solve for x and y. Confirm your answers by measuring the angles with a protractor.

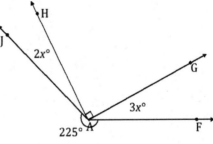

Example 3

In a complete sentence, describe the angle relationships in the diagram. Write an equation for the angle relationship shown in the figure and solve for x. Find the measures of $\angle JAH$ and $\angle GAF$. Confirm your answers by measuring the angles with a protractor.

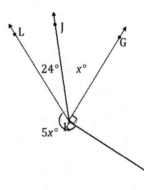

Exercise 3

In a complete sentence, describe the angle relationships in the diagram. Write an equation for the angle relationship shown in the figure and solve for x. Find the measure of $\angle JKG$. Confirm your answer by measuring the angle with a protractor.

Example 4

In the accompanying diagram, the measure of ∠DBE is four times the measure of ∠FBG.

    a.   Label ∠DBE as $y°$ and ∠FBG as $x°$. Write an equation that
          describes the relationship between ∠DBE and ∠FBG.

    b.   Find the value of x.

    c.   Find the measures of ∠FBG, ∠CBD, ∠ABF, ∠GBE, and ∠DBE.

    d.   What is the measure of ∠ABG? Identify the angle relationship used to get your answer.

Name _____   Date _____

Write an equation for the angle relationship shown in the figure and solve for x. Find the measures of $\angle RQS$ and $\angle TQU$.

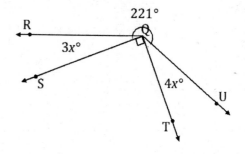

In a complete sentence, describe the angle relationships in each diagram. Write an equation for the angle relationship(s) shown in the figure, and solve for the indicated unknown angle.

1. Find the measure of $\angle HLG$.

    $\angle BLC$, $\angle CLD$, and $\angle DLE$ have a sum of $90°$.

    $\angle ALK$, $\angle KLJ$, $\angle JLH$, $\angle HLG$, and $\angle GLF$ are angles on a line and have a sum of $180°$.

$$3x + 12x + 30 = 90$$
$$15x + 30 = 90$$
$$15x + 30 - 30 = 90 - 30$$
$$15x = 60$$
$$\left(\frac{1}{15}\right)15x = \left(\frac{1}{15}\right)60$$
$$x = 4$$

I look for the small box in the corner showing me 90° angles, and I also look for angles that form a straight line because their measures have a sum of 180°.

In order to solve for y, I must solve for x first. The value of x can be used to help determine the value of y.

$$15x + 5x + 10x + y + 6x = 180$$
$$15(4) + 5(4) + 10(4) + y + 6(4) = 180$$
$$60 + 20 + 40 + y + 24 = 180$$
$$144 + y = 180$$
$$144 - 144 + y = 180 - 144$$
$$y = 36$$

$m\angle HLG = 36°$

2. Find the measures of ∠*TWV* and ∠*ZWV*.

The measures of ∠*XWY* and ∠*YWZ* have a sum of 90°.
The measures of ∠*TWV*, ∠*VWZ*, and ∠*ZWY* have a sum
of 180°.

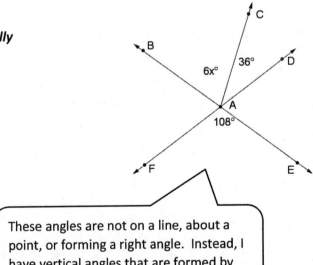

$$90 - 65 = 25$$
$$m∠ZWY = 25°$$

> I know that ∠ *XWY* and ∠ *ZWY* have a sum of 90°. I can work backwards to determine the unknown angle.

$$24x + 7x + 25 = 180$$
$$31x + 25 = 180$$
$$31x + 25 - 25 = 180 - 25$$
$$31x = 155$$
$$\left(\frac{1}{31}\right)31x = \left(\frac{1}{31}\right)155$$
$$x = 5$$

> I don't have all the information to use the angles on a line to solve for *x* yet, but I can get the measure of ∠*ZWY* by knowing that there are two angles that have a sum of 90° first.

$$m∠TWV = 24(5°) = 120°$$
$$m∠VWZ = 7(5°) = 35°$$

3. Find the measure of ∠*BAC*.

Adjacent angles 6*x*° and 36° together are vertically
opposite from and are equal to angle 108°.

$$6x + 36 = 108$$
$$6x + 36 - 36 = 108 - 36$$
$$6x = 72$$
$$\left(\frac{1}{6}\right)6x = \left(\frac{1}{6}\right)72$$
$$x = 12$$

$$m∠BAC = 6(12°) = 72°$$

> These angles are not on a line, about a point, or forming a right angle. Instead, I have vertical angles that are formed by two intersecting lines.

4. The measures of three angles at a point are in the ratio of 2 : 7 : 9. Find the measures of the angles.

> I can use the ratio to write expressions to represent each of the angles.

$m\angle A = 2x°, m\angle B = 7x°, m\angle C = 9x°$

$$2x + 7x + 9x = 360$$
$$18x = 360$$
$$\left(\frac{1}{18}\right)18x = \left(\frac{1}{18}\right)360$$
$$x = 20$$

> Since these three angles are at a point, their measures have a sum of 360°.

$m\angle A = 2(20°) = 40°$

$m\angle B = 7(20°) = 140°$

$m\angle C = 9(20°) = 180°$

In a complete sentence, describe the angle relationships in each diagram. Write an equation for the angle relationship(s) shown in the figure, and solve for the indicated unknown angle. You can check your answers by measuring each angle with a protractor.

1. Find the measures of ∠EAF, ∠DAE, and ∠CAD.

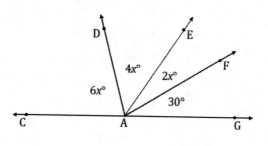

2. Find the measure of a.

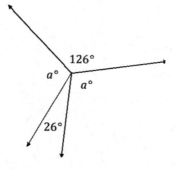

3. Find the measures of x and y.

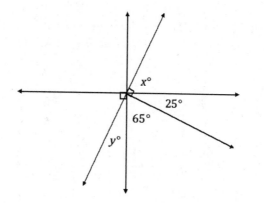

4. Find the measure of ∠HAJ.

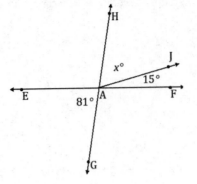

5. Find the measures of ∠HAB and ∠CAB.

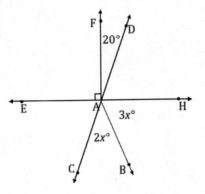

6. The measure of ∠SPT is b°. The measure of ∠TPR is five more than two times ∠SPT. The measure of ∠QPS is twelve less than eight times the measure of ∠SPT. Find the measures of ∠SPT, ∠TPR, and ∠QPS.

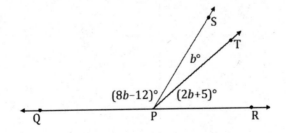

7. Find the measures of ∠*HQE* and ∠*AQG*.

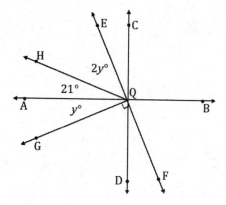

8. The measures of three angles at a point are in the ratio of 2 : 3 : 5. Find the measures of the angles.

9. The sum of the measures of two adjacent angles is 72°. The ratio of the smaller angle to the larger angle is 1 : 3. Find the measure of each angle.

10. Find the measures of ∠*CQA* and ∠*EQB*.

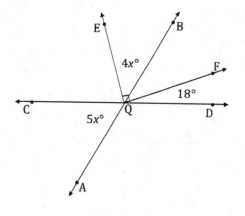

Example 1

Preserves the inequality symbol:

Reverses the inequality symbol:

Station 1

Die 1	Inequality	Die 2	Operation	New Inequality	Inequality Symbol Preserved or Reversed?
-3	$<$	5	Add 2	$-3 + 2 < 5 + 2$ $-1 < 7$	Preserved
			Add -3		
			Subtract 2		
			Subtract -1		
			Add 1		

Examine the results. Make a statement about what you notice, and justify it with evidence.

Station 2

Die 1	Inequality	Die 2	Operation	New Inequality	Inequality Symbol Preserved or Reversed?
-3	$<$	4	Multiply by -1	$(-1)(-3) < (-1)(4)$ $3 > -4$	Reversed
			Multiply by -1		
			Multiply by -1		
			Multiply by -1		
			Multiply by -1		

Examine the results. Make a statement about what you notice and justify it with evidence.

when negative multipliers can in place it changes whether the number is positive or negative

Lesson 12: Properties of Inequalities

Station 3

Die 1	Inequality	Die 2	Operation	New Inequality	Inequality Symbol Preserved or Reversed?
-2	$>$	-4	Multiply by $\dfrac{1}{2}$	$(-2)\left(\dfrac{1}{2}\right) > (-4)\left(\dfrac{1}{2}\right)$ $-1 > -2$	Preserved
			Multiply by 2		
			Divide by 2		
			Divide by $\dfrac{1}{2}$		
			Multiply by 3		

Examine the results. Make a statement about what you notice, and justify it with evidence.

Station 4

Die 1	Inequality	Die 2	Operation	New Inequality	Inequality Symbol Preserved or Reversed?
3	>	−2	Multiply by −2	$3(-2) > (-2)(-2)$ $-6 < 4$	Reversed
			Multiply by −3		
			Divide by −2		
			Divide by $-\dfrac{1}{2}$		
			Multiply by $-\dfrac{1}{2}$		

Examine the results. Make a statement about what you notice and justify it with evidence.

Exercise

Complete the following chart using the given inequality, and determine an operation in which the inequality symbol is preserved and an operation in which the inequality symbol is reversed. Explain why this occurs.

Inequality	Operation and New Inequality Which Preserves the Inequality Symbol	Operation and New Inequality Which Reverses the Inequality Symbol	Explanation
$2 < 5$			
$-4 > -6$			
$-1 \leq 2$			
$-2 + (-3) < -3 - 1$			

Lesson Summary

When both sides of an inequality are added or subtracted by a number, the inequality symbol stays the same, and the inequality symbol is said to be _____.

When both sides of an inequality are multiplied or divided by a positive number, the inequality symbol stays the same, and the inequality symbol is said to be _____.

When both sides of an inequality are multiplied or divided by a negative number, the inequality symbol switches from < to > or from > to <. The inequality symbol is _____.

Die Templates

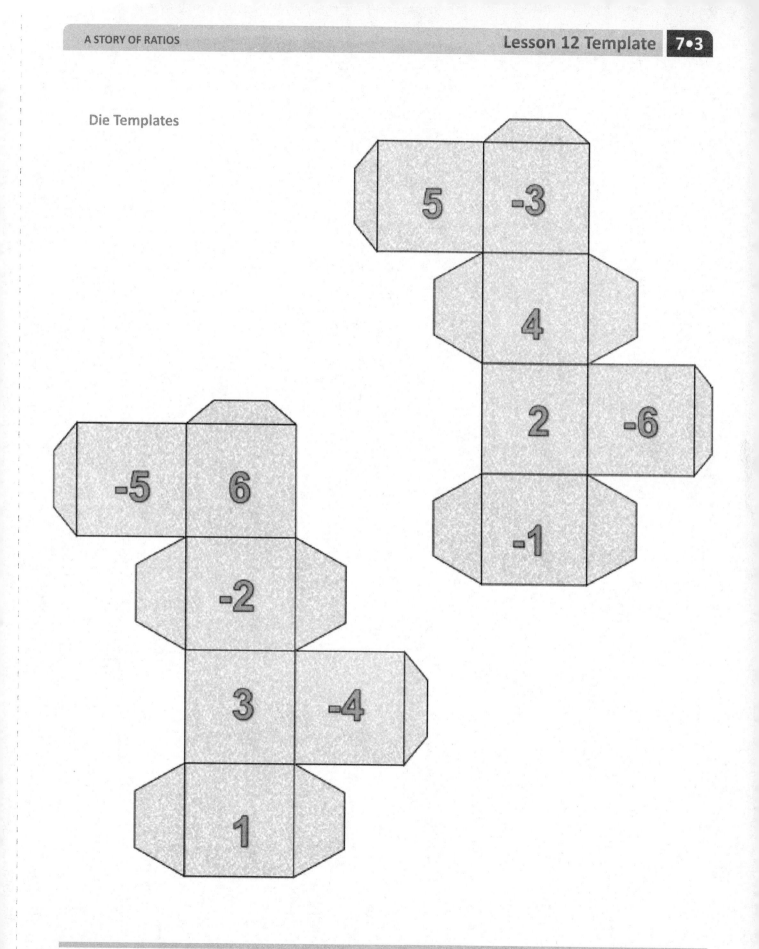

Name _____ Date _____

1. Given the initial inequality $-4 < 7$, state possible values for c that would satisfy the following inequalities.

    a. $c(-4) < c(7)$ $c = 2$

    $-6 < 14$

    b. $c(-4) > c(7)$ $c = -3$

    c. $c(-4) = c(7)$ $c = 0$

2. Given the initial inequality $2 > -4$, identify which operation preserves the inequality symbol and which operation reverses the inequality symbol. Write the new inequality after the operation is performed.

    a. Multiply both sides by -2.

    $-4 < 8$

    b. Add -2 to both sides.

    $0 > -6$

    c. Divide both sides by 2.

    $1 > -2$

    d. Multiply both sides by $-\frac{1}{2}$.

    $-1 < 2$

    e. Subtract -3 from both sides.

    $5 > -1$

© 2019 Great Minds®. eureka-math.org

1. For each problem, use the properties of inequalities to write a true inequality statement.
   The two integers are -8 and -3.

   a. Write a true inequality statement.

   $-8 < -3$

   > I can picture a number line to help me write the inequality. On a number line, -8 would be to the left of -3, which means it is less than -3.

   b. Add -4 to each side of the inequality. Write a true inequality statement.

   $-12 < -7$

   > I need to add $-8 + -4$ and $-3 + -4$ and then write another inequality. I can always look back at Module 2 for help working with signed numbers.

   c. Multiply each number in part (a) by -5. Write a true inequality statement.

   $40 > 15$

   > I need to multiply -8×-5 and -3×-5 and then write another inequality. I notice that I must reverse the inequality sign in order to write a true statement.

   d. Subtract $-c$ from each side of the inequality in part (a). Assume that c is a positive number. Write a true inequality statement.

   $-8 - (-c) < -3 - (-c)$
   $\quad -8 + c < -3 + c$

   > I know that adding or subtracting an integer from both sides of the inequality preserves the inequality sign.

e.   Divide each side of the inequality in part (a) by $-c,$ where c is positive. Write a true inequality statement.

$$\frac{-8}{-c} > \frac{-3}{-c}$$

$$\frac{8}{c} > \frac{3}{c}$$

> I know that when I divide by a negative, the inequality symbol is reversed.

2.   Kyla and Pedro went on vacation in northern Vermont during the winter. On Monday, the temperature was $-30°\text{F}$, and on Wednesday the temperature was $-8°\text{F}$.

a.   Write an inequality comparing the temperature on Monday and the temperature on Wednesday.

$$-30 < -8$$

> I need to compare these temperatures using less than or greater than.

b.   If the temperatures felt 12 degrees colder each day with the wind chill, write a new inequality to show the comparison of the temperatures they actually felt.

$$-42 < -20$$

> I could show the temperature with the wind chill by adding -12 to both sides.

c.   Was the inequality symbol preserved or reversed? Explain.

The inequality symbol was preserved because the number was added or subtracted from both sides of the inequality.

     Lesson 12:   Properties of Inequalities

1. For each problem, use the properties of inequalities to write a true inequality statement.
   The two integers are −2 and −5.

   a. Write a true inequality statement.

   b. Subtract −2 from each side of the inequality. Write a true inequality statement.

   c. Multiply each number by −3. Write a true inequality statement.

2. On a recent vacation to the Caribbean, Kay and Tony wanted to explore the ocean elements. One day they went in a submarine 150 feet below sea level. The second day they went scuba diving 75 feet below sea level.

   a. Write an inequality comparing the submarine's elevation and the scuba diving elevation.

   b. If they only were able to go one-fifth of the capable elevations, write a new inequality to show the elevations they actually achieved.

   c. Was the inequality symbol preserved or reversed? Explain.

3. If a is a negative integer, then which of the number sentences below is true? If the number sentence is not true, give a reason.

   a. $5 + a < 5$

   b. $5 + a > 5$

   c. $5 - a > 5$

   d. $5 - a < 5$

   e. $5a < 5$

   f. $5a > 5$

   g. $5 + a > a$

   h. $5 + a < a$

   i. $5 - a > a$

   j. $5 - a < a$

   k. $5a > a$

   l. $5a < a$

Opening Exercise: Writing Inequality Statements

Tarik is trying to save $265.49 to buy a new tablet. Right now, he has $40 and can save $38 a week from his allowance.

Write and evaluate an expression to represent the amount of money saved after …

2 weeks

$116   40 + (38·2) =

3 weeks

$154

4 weeks

$192

5 weeks

$230

6 weeks

$268

7 weeks

306

8 weeks

344

When will Tarik have enough money to buy the tablet?

Yes

Write an inequality that will generalize the problem.

$$40 + 38h \geq 265.44$$

Example 1: Evaluating Inequalities—Finding a Solution

The sum of two consecutive odd integers is more than −12. Write several true numerical inequality expressions.

J = odd numbers

2×2 > -12

atleast -14 and -15

x > -14

The sum of two consecutive odd integers is more than −12. What is the smallest value that will make this true?

a.   Write an inequality that can be used to find the smallest value that will make the statement true.

© 2019 Great Minds®. eureka-math.org

b. Use if-then moves to solve the inequality written in part (a). Identify where the 0's and 1's were made using the if-then moves.

c. What is the smallest value that will make this true?

Exercises

1. Connor went to the county fair with $22.50 in his pocket. He bought a hot dog and drink for $3.75 and then wanted to spend the rest of his money on ride tickets, which cost $1.25 each.

    a. Write an inequality to represent the total spent where r is the number of tickets purchased.

$$1.25r + 3.75 \leq 22.50$$

    b. Connor wants to use this inequality to determine whether he can purchase 10 tickets. Use substitution to show whether he will have enough money.

$$12.5 + 3.75 = 16.25$$

$$16.25 \leq 22.5$$

✓

c.   What is the total maximum number of tickets he can buy based upon the given information?

$$1.25r + 3.75 = 22.50$$
$$ -3.75 -3.75$$
$$\frac{1.25 r}{1.25} = \frac{18.75}{1.25}$$
$$r = 15$$

15 tickets is max

2.   Write and solve an inequality statement to represent the following problem:

On a particular airline, checked bags can weigh no more than 50 pounds. Sally packed 32 pounds of clothes and five identical gifts in a suitcase that weighs 8 pounds. Write an inequality to represent this situation.

$$5_g + 8 + 32 \le 50$$

$$5_g + 40 \le 50$$

$$g \le 2$$

Lesson 13:   Inequalities

Name _____   Date _____

Shaggy earned \$7.55 per hour plus an additional \$100 in tips waiting tables on Saturday. He earned at least \$160 in all. Write an inequality and find the minimum number of hours, to the nearest hour, that Shaggy worked on Saturday.

$$7.55h + 100 \geq 160$$
$$\underline{-100} \qquad \underline{-100}$$
$$0 \qquad\qquad$$
$$7.55h \geq 60$$
$$h \geq 7$$

I notice that the problem states that x is a positive integer, which means that x could be $1, 2, 3, 4, 5, 6, \ldots$.

1. If x represents a positive integer, find the solutions to the following inequalities.

   a. $x + 9 \leq 5$

$$x + 9 \leq 5$$
$$x + 9 - 9 \leq 5 - 9$$
$$x \leq -4$$

I determined that the only values of x that will make the inequality true are less than or equal to -4, but there are no positive integers that are less than or equal to -4.

There are no positive integers that are a solution.

   b. $5 + \dfrac{x}{7} > 12$

I can solve inequalities similar to how I solve equations, but I remember that there are times when I have to reverse the inequality symbol.

$$5 - 5 + \dfrac{x}{7} > 12 - 5$$
$$\dfrac{x}{7} > 7$$
$$7\left(\dfrac{x}{7}\right) > 7\,(7)$$
$$x > 49$$

The possible solutions for x would include all integers greater than 49.

If x is greater than 49, then it cannot be exactly 49. Instead, x may be $50, 51, 52, 53, \ldots$ or any larger integer.

Lesson 13: Inequalities

For each part, I need to determine if any negative integer will be a possible solution, if only some negative integers are solutions, or if no negative integer could ever be a solution.

2. Recall that the symbol ≠ means not equal to. If x represents a negative integer, state whether each of the following statements is always true, sometimes true, or false.

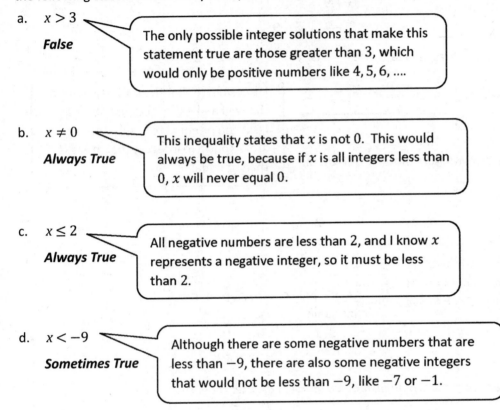

a. $x > 3$

False

The only possible integer solutions that make this statement true are those greater than 3, which would only be positive numbers like $4, 5, 6, ...$

b. $x \neq 0$

Always True

This inequality states that x is not 0. This would always be true, because if x is all integers less than 0, x will never equal 0.

c. $x \leq 2$

Always True

All negative numbers are less than 2, and I know x represents a negative integer, so it must be less than 2.

d. $x < -9$

Sometimes True

Although there are some negative numbers that are less than -9, there are also some negative integers that would not be less than -9, like -7 or -1.

EUREKA MATH

3. Three times the smaller of two consecutive even integers increased by the larger integer is at least 26.

> Consecutive even integers are two apart, so I use x to represent the first number and $x + 2$ to represent the second number.

Model the problem with an inequality, and determine which of the given values 4 and 6 are solutions. Then, find the smallest number that will make the inequality true.

$3x + x + 2 \geq 26$

> I know that "at least" means that the sum will be 26 or more. So the sum will be greater than or equal to 26.

For 4:

$3(4) + (4) + 2 \geq 26$

$12 + 4 + 2 \geq 26$

$18 \geq 26$

For 6:

$3(6) + (6) + 2 \geq 26$

$18 + 6 + 2 \geq 26$

$26 \geq 26$

False, 4 is not a solution.

True, 6 is a solution.

> To determine if a number is a solution of an inequality, I can just substitute the number in for x and evaluate to see if the result is a true statement.

$3x + x + 2 \geq 26$

$4x + 2 \geq 26$

$4x + 2 - 2 \geq 26 - 2$

$4x \geq 24$

$x \geq 6$

The smallest number that will make the inequality true is 6.

> "At most" tells me that she can use 74 feet of fencing or less. So the total must be less than or equal to 74.

4. Rochelle has, at most, 74 feet of fencing to put around her veggie garden. She plans to create a rectangular garden that has a length that is 3 feet longer than the width. Write an inequality to model the situation. Then solve to determine the dimensions of the garden with the largest perimeter Rochelle can make.

   Let x represent the width.

   Let x + 3 represent the length.

   > The perimeter is the sum of all 4 sides of a rectangle. I must include all 4 sides when writing my inequality.

$$x + x + x + 3 + x + 3 \leq 74$$
$$4x + 6 \leq 74$$
$$4x + 6 - 6 \leq 74 - 6$$
$$4x \leq 68$$
$$\left(\frac{1}{4}\right)(4x) \leq \left(\frac{1}{4}\right)(68)$$
$$x \leq 17$$

$$17 + 3 = 20$$

In order to get the largest perimeter, the width would be 17 feet, and the length would be 20 feet.

> Once I solve, I need to use the "let" statements I wrote at the beginning to help me determine the meaning of my answer.

1. Match each problem to the inequality that models it. One choice will be used twice.

   _____c_____ The sum of three times a number and −4 is greater than 17.

   _____b_____ The sum of three times a number and −4 is less than 17.

   _____d_____ The sum of three times a number and −4 is at most 17.

   _____a_____ The sum of three times a number and −4 is no more than 17.

   _____a_____ The sum of three times a number and −4 is at least 17.

   a. $3x + -4 \geq 17$

   b. $3x + -4 < 17$

   c. $3x + -4 > 17$

   d. $3x + -4 \leq 17$

2. If x represents a positive integer, find the solutions to the following inequalities.

   a. $x < 7$

   b. $x - 15 < 20$

   c. $x + 3 \leq 15$

   d. $-x > 2$

   e. $10 - x > 2$

   f. $-x \geq 2$

   g. $\dfrac{x}{3} < 2$

   h. $-\dfrac{x}{3} > 2$

   i. $3 - \dfrac{x}{4} > 2$

3. Recall that the symbol $\neq$ means *not equal to*. If x represents a positive integer, state whether each of the following statements is always true, sometimes true, or false.

   a. $x > 0$

   b. $x < 0$

   c. $x > -5$

   d. $x > 1$

   e. $x \geq 1$

   f. $x \neq 0$

   g. $x \neq -1$

   h. $x \neq 5$

4. Twice the smaller of two consecutive integers increased by the larger integer is at least 25.

   Model the problem with an inequality, and determine which of the given values 7, 8, and/or 9 are solutions. Then, find the smallest number that will make the inequality true.

5.
   a. The length of a rectangular fenced enclosure is 12 feet more than the width. If Farmer Dan has 100 feet of fencing, write an inequality to find the dimensions of the rectangle with the largest perimeter that can be created using 100 feet of fencing. $(x+12)+(x)+(x+12)+(x) \leq 100$

   b. What are the dimensions of the rectangle with the largest perimeter? What is the area enclosed by this rectangle

   $3x + 24 = 100$
   $\dfrac{-24}{3x} = \dfrac{-24}{76}$ $x = 25\frac{1}{3}$ $l = 27\frac{1}{3}$

6. At most, Kyle can spend $50 on sandwiches and chips for a picnic. He already bought chips for $6 and will buy sandwiches that cost $4.50 each. Write and solve an inequality to show how many sandwiches he can buy. Show your work and interpret your solution.

   $4.50x + 6 \leq 50$
   $4.50x \leq \dfrac{-6}{46}$ 9 sandwiches

© 2019 Great Minds®. eureka-math.org

$6P - 265 \geq 1000$

$6P \geq 1265$

$P \geq$

$\boxed{211}$

$1266 - 200 \cdot 5$

$\Rightarrow 1100$

$$\begin{array}{r} 40.8 \\ 6\overline{\smash{)}1265} \\ 18 \\ \hline 6 \\ 6 \\ \hline 50 \end{array}$$

Opening Exercise

The annual County Carnival is being held this summer and will last $5\frac{1}{2}$ days. Use this information and the other given information to answer each problem.

You are the owner of the biggest and newest roller coaster called the Gentle Giant. The roller coaster costs $6 to ride. The operator of the ride must pay $200 per day for the ride rental and $65 per day for a safety inspection. If you want to make a profit of at least $1,000 each day, what is the minimum number of people that must ride the roller coaster?

Write an inequality that can be used to find the minimum number of people, p, which must ride the roller coaster each day to make the daily profit.

$$6p - (100 + 357.5) \geq 1,000$$

$$6p - 1457.5 \geq 1000$$

Solve the inequality.

$$6p - 1457.5 \geq 1,000$$

$$+1,457.5$$

$$6 \div 6p = 2,457.5 \div 6$$

$$p = 413$$

$$413 \div 5\frac{1}{2}$$

$$6\overline{)2,475.5} \quad 412.583$$

Interpret the solution.

413 people need to ride to earn $1,000

Example 1

A youth summer camp has budgeted $2,000 for the campers to attend the carnival. The cost for each camper is $17.95, which includes general admission to the carnival and two meals. The youth summer camp must also pay $250 for the chaperones to attend the carnival and $350 for transportation to and from the carnival. What is the greatest number of campers who can attend the carnival if the camp must stay within its budgeted amount?

Example 2

The carnival owner pays the owner of an exotic animal exhibit $650 for the entire time the exhibit is displayed. The owner of the exhibit has no other expenses except for a daily insurance cost. If the owner of the animal exhibit wants to make more than $500 in profits for the $5\frac{1}{2}$ days, what is the greatest daily insurance rate he can afford to pay?

Example 3

Several vendors at the carnival sell products and advertise their businesses. Shane works for a recreational company that sells ATVs, dirt bikes, snowmobiles, and motorcycles. His boss paid him $500 for working all of the days at the carnival plus 5% commission on all of the sales made at the carnival. What was the minimum amount of sales Shane needed to make if he earned more than $1,500?

Lesson Summary

The key to solving inequalities is to use if-then moves to make 0's and 1's to get the inequality into the form $x > c$ or $x < c$ where c is a number. Adding or subtracting opposites will make 0's. According to the if-then move, any number that is added to or subtracted from each side of an inequality does not change the solution to the inequality. Multiplying and dividing numbers makes 1's. When each side of an inequality is multiplied by or divided by a positive number, the sign of the inequality is not reversed. However, when each side of an inequality is multiplied by or divided by a negative number, the sign of the inequality is reversed.

Given inequalities containing decimals, equivalent inequalities can be created which have only integer coefficients and constant terms by repeatedly multiplying every term by ten until all coefficients and constant terms are integers.

Given inequalities containing fractions, equivalent inequalities can be created which have only integer coefficients and constant terms by multiplying every term by the least common multiple of the values in the denominators.

Name _____   Date _____

Games at the carnival cost $3 each. The prizes awarded to winners cost the owner $145.65. How many games must be played for the owner to make at least $50? $g =$ games

$$3g - 145.65 \geq 50$$
$$+ 145.65 \quad\quad + 145.65$$

$$3 \div 3g \geq 195.65 \div 3$$

$$g \geq 65$$

69 games
for owner to make
at least $50

1. Ethan earns a commission of 5% of the total amount he sells. In addition, he is also paid $380 per week. In order to stick to his budget, he needs to earn at least $975 this week. Write an inequality with integer coefficients for the total sales needed to earn at least $975, and describe what the solution represents.

> Because he has to earn at least $975, I know that I should use greater than or equal to 975 because Ethan needs to earn $975 or more.

Let the variable p represent the purchase amount.

> Since percent means out of 100, I can show 5% as $\frac{5}{100}$.

$$\frac{5}{100}p + 380 \geq 975$$

$$(100)\left(\frac{5}{100}p\right) + 100(380) \geq 100(975)$$

> Now that I have gotten rid of the fraction, this would be my inequality with integer coefficients.

$$5p + 38000 \geq 97500$$
$$5p + 38000 - 38000 \geq 97500 - 38000$$
$$5p \geq 59500$$
$$\left(\frac{1}{5}\right)(5p) \geq \left(\frac{1}{5}\right)(59500)$$
$$p \geq 11900$$

Ethan's total sales must be at least $11,900 if he wants to earn $975 or more.

2. Katie and Kane were exercising on Saturday. Kane was riding a bicycle 12 miles per hour faster than Katie was walking. Katie walked for $3\frac{1}{2}$ hours, and Kane bicycled for 2 hours. Altogether, Katie and Kane traveled no more than 57 miles. Find the maximum speed of each person.

	Rate	Time	Distance
Kane	$x + 12$	2	$2(x + 12)$
Katie	x	$3\frac{1}{2}$	$3\frac{1}{2}x$

> I can organize all the information in a table and use the relationship $d = rt$.

$$2(x + 12) + 3\frac{1}{2}x \leq 57$$

I know that sum of the two distances presented in the table must be no more than 57 miles, which means the sum must be less than or equal to 57.

$$2x + 24 + 3\frac{1}{2}x \leq 57$$

$$5\frac{1}{2}x + 24 \leq 57$$

$$5\frac{1}{2}x + 24 - 24 \leq 57 - 24$$

$$5\frac{1}{2}x \leq 33$$

Rewriting mixed numbers as fractions greater than one can help make it easier to solve.

$$\frac{11}{2}x \leq 33$$

$$\left(\frac{2}{11}\right)\left(\frac{11}{2}x\right) \leq (33)\left(\frac{2}{11}\right)$$

$$x \leq 6$$

$$6 + 12 = 18$$

The maximum speed Katie was walking was 6 miles per hour, and the maximum speed Kane was riding the bike was 18 miles per hour.

3. Systolic blood pressure is the higher number in a blood pressure reading. It is measured as the heart muscle contracts. Ramel is having his blood pressure checked. The nurse told him that the upper limit of his systolic blood pressure is equal to a third of his age increased by 117. If Ramel is 42 years old, write and solve an inequality to determine what is normal for his systolic blood pressure.

An upper limit provides the maximum number for Ramel's systolic blood pressure. That means that Ramel's systolic blood pressure must be less than or equal to the expression showing a third of his age increased by 117.

Let p represent the systolic blood pressure in millimeters of mercury (mmHg).

Let a represent Ramel's age.

$$p \leq \frac{1}{3}a + 117, \text{ where } a = 42.$$

$$p \leq \frac{1}{3}(42) + 117$$

I can just substitute in Ramel's age for a in order to solve for p.

$$p \leq 14 + 117$$

$$p \leq 131$$

The normal upper limit for Ramel is 131, which means that Ramel's systolic blood pressure should be 131 mmHg or lower.

1. As a salesperson, Jonathan is paid $50 per week plus 3% of the total amount he sells. This week, he wants to earn at least $100. Write an inequality with integer coefficients for the total sales needed to earn at least $100, and describe what the solution represents. $s = sales$

   1667 sales to make 100 $0.03s + 50 \geq 100$ $1.03 \equiv 50$ $s \leq 1666.67$

2. Systolic blood pressure is the higher number in a blood pressure reading. It is measured as the heart muscle contracts. Heather was with her grandfather when he had his blood pressure checked. The nurse told him that the upper limit of his systolic blood pressure is equal to half his age increased by 110.

   a. a is the age in years, and p is the systolic blood pressure in millimeters of mercury (mmHg). Write an inequality to represent this situation. $0.5a + 110 = P$

   b. Heather's grandfather is 76 years old. What is *normal* for his systolic blood pressure?

   148 mmHg

3. Traci collects donations for a dance marathon. One group of sponsors will donate a total of $6 for each hour she dances. Another group of sponsors will donate $75 no matter how long she dances. What number of hours, to the nearest minute, should Traci dance if she wants to raise at least $1,000? $h = hours$

   154 hours $6h + 75 \geq 1,000$ $6h = 925$ $h \leq 154.17$

4. Jack's age is three years more than twice the age of his younger brother, Jimmy. If the sum of their ages is at most 18, find the greatest age that Jimmy could be. $x = 3 + 2y$

   $3y + 3 \leq 18$ $3y = 15$ $y = 5$ $x = 13$

5. Brenda has $500 in her bank account. Every week she withdraws $40 for miscellaneous expenses. How many weeks can she withdraw the money if she wants to maintain a balance of a least $200? $w = weeks$

   7.5 weeks $500 - 40w \geq 200$ $-40w \geq -300$ $w = 7.5$

6. A scooter travels 10 miles per hour faster than an electric bicycle. The scooter traveled for 3 hours, and the bicycle traveled for $5\frac{1}{2}$ hours. Altogether, the scooter and bicycle traveled no more than 285 miles. Find the maximum speed of each. $s = $ bs

$$10 + 2b + 8\frac{1}{2} \leq 285$$

$$18\frac{1}{2} + 2b \leq 285$$

$(10 + b) + b$ -11.5

$(10 + 2b)\, 8\frac{1}{2} \leq 285$

$(10 + b) \times 3 + b \times 5\frac{1}{2} \leq 285$

$30 + 3b + 5\frac{1}{2}b \leq 285$

$s = \dfrac{d}{t}$ $(s \wedge t) = d$ $8\frac{1}{2}b \leq 255$

$b \leq \dfrac{255}{8.5}$

$b \leq \dfrac{255}{30}$ $s = 40$

Exercise 1

1. Two identical cars need to fit into a small garage. The opening is 23 feet 6 inches wide, and there must be at least 3 feet 6 inches of clearance between the cars and between the edges of the garage. How wide can the cars be?

$$2x + 10.5 \leq 23.5$$
$$ {-10.5}$$
$$\div 2 \quad 2x \leq 13.2$$
$$x \leq 6.5$$

Cars cane be
6.5" wide

23.5 ft

Example

A local car dealership is trying to sell all of the cars that are on the lot. Currently, there are 525 cars on the lot, and the general manager estimates that they will consistently sell 50 cars per week. Estimate how many weeks it will take for the number of cars on the lot to be less than 75.

Write an inequality that can be used to find the number of full weeks, w, it will take for the number of cars to be less than 75. Since w is the number of full or complete weeks, $w = 1$ means at the end of week 1.

$$525 - 50w < 75$$
$$-525 \qquad -525$$
$$-50w < -450 \div -50$$

Solve and graph the inequality.

$$w > 9$$

Interpret the solution in the context of the problem.

The dealership can have less than 75 cars remaining if they sell 50 cars per week for more than 9 weeks

Verify the solution.

$$v = 10$$

$$50(10)$$

$$525 - 500 < 75$$

$$25 < 75 \quad \text{True}$$

Exercise 2

2. The cost of renting a car is $25 per day plus a one-time fee of $75.50 for insurance. How many days can the car be rented if the total cost is to be no more than $525? $d = days$

   a. Write an inequality to model the situation.

   $$25d + 75.50 \le 525$$

   b. Solve and graph the inequality.

   $$25d + 75.50 \le 525$$
   $$\underline{-75.50 \quad -75.5}$$
   $$25 \div 25d \le 449.5 \div 25$$
   $$d \le 17.18$$

   c. Interpret the solution in the context of the problem.

   The car can be rented for 17 or less days within $525. 18th day would make fee over $525. Days are pos

       Lesson 15:   Graphing Solutions to Inequalities

EUREKA MATH

Additional Exercises

For each problem, write, solve, and graph the inequality, and interpret the solution within the context of the problem.

3. Mrs. Smith decides to buy three sweaters and a pair of jeans. She has $120 in her wallet. If the price of the jeans is $35, what is the highest possible price of a sweater, if each sweater is the same price? $S = $ sweater

$$3s + 35 \leq 120$$
$$-35 \quad -35$$
$$3 \div 3s \leq \frac{85}{3} \cdot 3$$
$$s \leq 28.\overline{3}$$

26 27 28 29 30

Highest price is $28.33

4. The members of the Select Chorus agree to buy at least 250 tickets for an outside concert. They buy 20 fewer lawn tickets than balcony tickets. What is the least number of balcony tickets bought?

$b = $ balcony
$l = $ lawn $= b - 20$

$$(b-20) + b \geq 250$$
$$2b - 20 \geq 250$$
$$+20 \qquad +20$$
$$2b = 270 \div 2$$
$$b = 135$$
$$l = 115$$

least number of balcony tickets bought is 135.

133 134 135 136 137

© 2019 Great Minds®. eureka-math.org

5. Samuel needs $29 to download some songs and movies on his MP3 player. His mother agrees to pay him $6 an hour for raking leaves in addition to his $5 weekly allowance. What is the minimum number of hours Samuel must work in one week to have enough money to purchase the songs and movies?   h s hours

$$6h + 5 \geq 29$$
$$-5 \quad -25$$
$$6: 6h \geq \frac{24 \div 6}{4}$$
$$h \geq$$

The minimum number of hours Samuel must work to achieve enough money is 4.

EUREKA MATH®

Name _____ Date _____

The junior high art club sells candles for a fundraiser. The first week of the fundraiser, the club sells 7 cases of candles. Each case contains 40 candles. The goal is to sell at least 13 cases. During the second week of the fundraiser, the club meets its goal. Write, solve, and graph an inequality that can be used to find the possible number of candles sold the second week.

$c \leq$ can

$$7(40) + 40c \leq 13$$

$$280 + 40c \leq 13$$

$$\frac{-28}{40c = -267}$$

240
candles
on 2nd week

$$280 + c \leq 520$$

$$\frac{-280}{c \leq 240}$$

13
×40
520

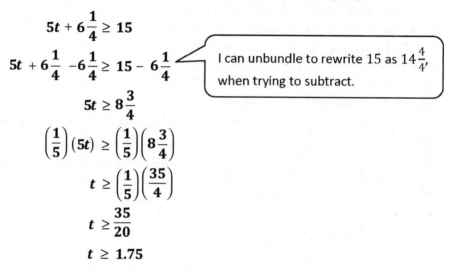

Because this problem says at least 15 hours, I know that he must read 15 hours or more. I use the greater than or equal to symbol in my inequality to represent this relationship.

1. Doug has decided that he should read for at least 15 hours a week. On Monday and Tuesday, his days off from work, he reads for a total of $6\frac{1}{4}$ hours. For the remaining 5 days, he reads for the same amount of time each day. Find t, the amount of time he reads for each of the 5 days. Graph your solution.

   Let t represent the time, in hours, he spends reading on each of the remaining days.

$$5t + 6\frac{1}{4} \geq 15$$

$$5t + 6\frac{1}{4} - 6\frac{1}{4} \geq 15 - 6\frac{1}{4}$$

I can unbundle to rewrite 15 as $14\frac{4}{4}$, when trying to subtract.

$$5t \geq 8\frac{3}{4}$$

$$\left(\frac{1}{5}\right)(5t) \geq \left(\frac{1}{5}\right)\left(8\frac{3}{4}\right)$$

$$t \geq \left(\frac{1}{5}\right)\left(\frac{35}{4}\right)$$

$$t \geq \frac{35}{20}$$

$$t \geq 1.75$$

Doug must read for 1.75 hours or more on each of the remaining days.

Graph:

Because I want to include 1.75 as a possible solution, I use a solid circle. The arrow indicates that all numbers greater than 1.75 are also included in the solution.

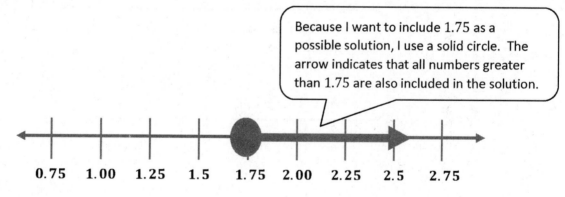

2. The length of a parallelogram is 70 centimeters, and its perimeter is less than 360 centimeters. Cherise writes an inequality and graphs the solution below to find the width of the parallelogram. Is she correct? If yes, write and solve the inequality to represent the problem and graph. If no, explain the error(s) Cherise made.

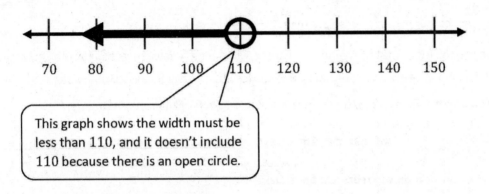

This graph shows the width must be less than 110, and it doesn't include 110 because there is an open circle.

Let w represent the width of the parallelogram.

$$2w + 2\,(70) < 360$$
$$2w + 140 < 360$$
$$2w + 140 - 140 < 360 - 140$$
$$2w < 220$$
$$\left(\frac{1}{2}\right)(2w) < \left(\frac{1}{2}\right)(220)$$
$$w < 110$$

A parallelogram has two lengths and two widths. The sum of all four sides will be the perimeter. I can use this information to set up my inequality.

Yes, Cherise is correct.

The width must be less than 110 in order for the perimeter to be less than 360. To graph this relationship, I do need an open circle because 110 is not included in the solution. This matches the graph.

1. Ben has agreed to play fewer video games and spend more time studying. He has agreed to play less than 10 hours of video games each week. On Monday through Thursday, he plays video games for a total of $5\frac{1}{2}$ hours. For the remaining 3 days, he plays video games for the same amount of time each day. Find t, the amount of time he plays video games, for each of the 3 days. Graph your solution.

    Ben plays less than 1.5 hr for 3 of days

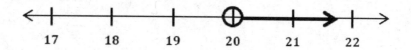

2. Gary's contract states that he must work more than 20 hours per week. The graph below represents the number of hours he can work in a week.

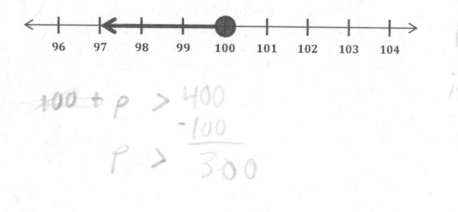

    a. Write an algebraic inequality that represents the number of hours, h, Gary can work in a week. $21h > 20$

    b. Gary is paid \$15.50 per hour in addition to a weekly salary of \$50. This week he wants to earn more than \$400. Write an inequality to represent this situation. $15.5h + 50 > 400$ (23)

    c. Solve and graph the solution from part (b). Round to the nearest hour.

    greater than (23)

3. Sally's bank account has \$650 in it. Every week, Sally withdraws \$50 to pay for her dog sitter. What is the maximum number of weeks that Sally can withdraw the money so there is at least \$75 remaining in the account? Write and solve an inequality to find the solution, and graph the solution on a number line.

4. On a cruise ship, there are two options for an Internet connection. The first option is a fee of \$5 plus an additional \$0.25 per minute. The second option costs \$50 for an unlimited number of minutes. For how many minutes, m, is the first option cheaper than the second option? Graph the solution.

    179 mins

    $0.25m + 5 < 50$

5. The length of a rectangle is 100 centimeters, and its perimeter is greater than 400 centimeters. Henry writes an inequality and graphs the solution below to find the width of the rectangle. Is he correct? If yes, write and solve the inequality to represent the problem and graph. If no, explain the error(s) Henry made.

    Henry is wrong it should be greater

    $100 + p > 400$
    -100
    $\overline{\;\;}$
    $p > 300$

$$3t + 5.5 < 10$$
$$-5.5$$
$$3 \div 3t = \frac{4.5 \div 3}{}$$
$$t = 1.5$$

Opening Exercise

   a.   Using a compass, draw a circle like the picture to the right.

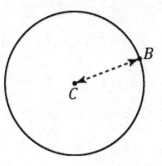

C is the *center* of the circle.
The distance between *C* and *B* is the *radius* of the circle.

   b.   Write your own definition for the term *circle*.

A round shape

   c.   Extend segment *CB* to a segment *AB* in part (a), where *A* is also a point on the circle.

The length of the segment *AB* is called the diameter of the circle.

   d.   The diameter is _____ 2 times _____ as long as the radius.

Lesson 16:   The Most Famous Ratio of All   207

e. Measure the radius and diameter of each circle. The center of each circle is labeled C.

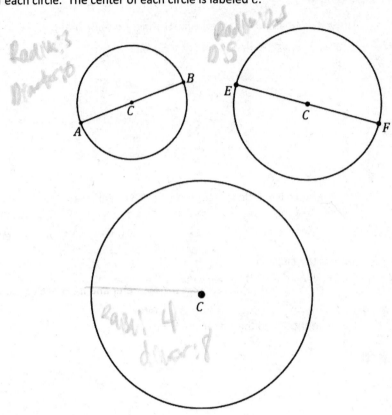

Radius: 3
Diameter: 6

Radius: 12nd
D'S

Radius: 4
diam: 8

f. Draw a circle of radius 6 cm.

Lesson 16:   The Most Famous Ratio of All

Mathematical Modeling Exercise

The ratio of the circumference to its diameter is always the same for any circle. The value of this ratio, $\frac{Circumference}{Diameter}$, is called the number *pi* and is represented by the symbol π.

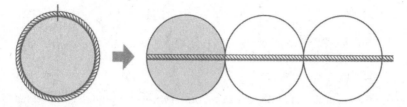

Since the circumference is a little greater than 3 times the diameter, π is a number that is a little greater than 3. Use the symbol π to represent this special number. Pi is a non-terminating, non-repeating decimal, and mathematicians use the symbol π or approximate representations as more convenient ways to represent pi.

- $\pi \approx 3.14$ or $\frac{22}{7}$.
- The ratios of the circumference to the diameter and $\pi : 1$ are equal.
- Circumference of a Circle $= \pi \times$ Diameter.

Example

a. The following circles are not drawn to scale. Find the circumference of each circle. (Use $\frac{22}{7}$ as an approximation for π.)

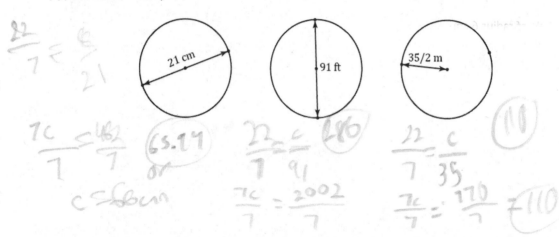

b. The radius of a paper plate is 11.7 cm. Find the circumference to the nearest tenth. (Use 3.14 as an approximation for π.)

diameter = 23.4   3.14 × 23.4 =   c = 73.476

C ≈ 73.5 cm

c.   The radius of a paper plate is 11.7 cm. Find the circumference to the nearest hundredth. (Use the π button on your calculator as an approximation for π.)

$$\frac{\pi}{1} = \frac{C}{23.4}$$

$$C = 23.4\pi$$

$$C = 73.51 \text{ cm}$$

d.   A circle has a radius of r cm and a circumference of C cm. Write a formula that expresses the value of C in terms of r and π.

$$2r \cdot \pi = C$$

e.   The figure below is in the shape of a semicircle. A semicircle is an arc that is half of a circle. Find the perimeter of the shape. (Use 3.14 for π.)

$$C = \pi \cdot 8$$

$$3.14 \cdot 8$$

8 m

$$\tfrac{1}{2}C = 25.12 \div 2 = \boxed{12.56 \text{ m}}$$

Lesson 16:   The Most Famous Ratio of All

Relevant Vocabulary

CIRCLE: Given a point O in the plane and a number $r > 0$, the *circle with center O and radius r* is the set of all points in the plane whose distance from the point O is equal to r.

RADIUS OF A CIRCLE: The *radius* is the length of any segment whose endpoints are the center of a circle and a point that lies on the circle.

DIAMETER OF A CIRCLE: The *diameter of a circle* is the length of any segment that passes through the center of a circle whose endpoints lie on the circle. If r is the *radius* of a circle, then the diameter is $2r$.

The word *diameter can* also mean the segment itself. Context determines how the term is being used: *The diameter* usually refers to the length of the segment, while *a diameter* usually refers to a segment. Similarly, *a radius* can refer to a segment from the center of a circle to a point on the circle.

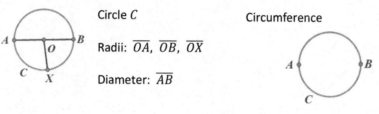

Circle C

Radii: $\overline{OA}$, $\overline{OB}$, $\overline{OX}$

Diameter: $\overline{AB}$

Circumference

CIRCUMFERENCE: The circumference of a circle is the distance around a circle.

PI: The number *pi,* denoted by π, is the value of the ratio given by the circumference to the diameter, that is $\pi = \dfrac{\text{circumference}}{\text{diameter}}$. The most commonly used approximations for π is 3.14 or $\dfrac{22}{7}$.

SEMICIRCLE: Let C be a circle with center O, and let A and B be the endpoints of a diameter. A *semicircle* is the set containing A, B, and all points that lie in a given half-plane determined by $\overline{AB}$ (diameter) that lie on circle C.

Semicircle

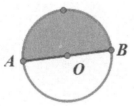

Name _____ Date _____

Brianna's parents built a swimming pool in the backyard. Brianna says that the distance around the pool is 120 feet.

1   Is she correct? Explain why or why not.

40 ft

20 ft

2.   Explain how Brianna would determine the distance around the pool so that her parents would know how many feet of stone to buy for the edging around the pool.

3.   Explain the relationship between the circumference of the semicircular part of the pool and the width of the pool.

1. Find the circumference.

   a. Give the exact answer in terms of π.

   $d = 10$ in.

   $C = \pi d$

   $C = 10\pi$ in.

   > I know that the diameter of every circle is twice the length of the radius, which is given in the diagram.

   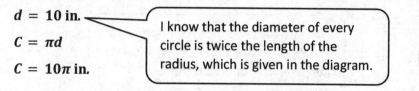

   b. Use $\pi \approx \frac{22}{7}$, and express your answer as a fraction in lowest terms.

   $C \approx \frac{22}{7}(10 \text{ in.})$

   $C \approx 31\frac{3}{7}$ in.

   > I use that the formula for circumference ($C = \pi d$) and use $\frac{22}{7}$ as the approximation for π.

   c. Use the π button on your calculator, and express your answer to the nearest hundredth.

   $C = \pi (10 \text{ in.})$

   $C \approx 31.42$ in.

   > In my calculator I type 10 and then press the multiplication and π buttons to calculate the circumference of the given circle.

2. Consider the diagram shown.

   a. Explain in words how to determine the perimeter of the diagram.

   The perimeter would be the sum of two side lengths (a) and (c) and the circumference of half a circle with diameter d.

   > To calculate the circumference of a half circle, I use the formula $C = \frac{1}{2}\pi d$.

   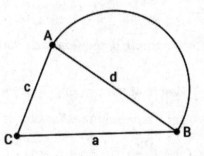

   b. Write an algebraic equation that will result in the perimeter of the diagram.

   $P = \frac{1}{2}\pi d + a + c$

   > I know the perimeter is the length around the outside of the diagram. It can be used to determine the amount of fencing or edging needed.

Lesson 16: The Most Famous Ratio of All

c. Find the perimeter of the figure if the diameter of the semicircle is 9 m, side length c is 6 m, and side length a is 10 m. Use 3.14 for π.

$$P = \frac{1}{2}\pi d + a + c$$

$$P \approx \frac{1}{2}(3.14)(9) + 10 + 6$$

I can use the given side lengths and the equation I wrote in part (b) to calculate the perimeter of the diagram.

$$P \approx 14.13 + 10 + 6$$

$$P \approx 30.13$$

The perimeter of the diagram is about 30.13 m.

3. Dan wants to go on the longest bike ride possible. If he plans to complete a loop where he starts at point D and ends back at point D, which route is the longest: following the semicircle path or following the path of the two smaller semicircles? Explain your reasoning. Let $\pi \approx 3.14$.

Length of the single semicircle path:

Let d represent the diameter of the semicircle.

$$P = \frac{1}{2}\pi d + d$$

In order to find the length of this entire path, I add the circumference of the semicircle and the length of the diameter.

$$P \approx \frac{1}{2}(3.14)(12) +$$

$$P \approx 18.84 + 12$$

$$P \approx 30.84$$

The length of the semicircle path is about 30.84 km.

If Dan was only riding his bike from point D to point E, he could either travel around the semicircle or follow the diameter of the circle.

Length of the two smaller semicircles path:

Let a represent the diameter of the smaller semicircles.

$$P = \frac{1}{2}\pi a + \frac{1}{2}\pi a + a + a$$

$$P \approx \frac{1}{2}(3.14)(6) + \frac{1}{2}(3.14)(6) + 6 + 6$$

The two smaller semicircles each have a diameter of 6 km. I also have to add the diameter of each smaller semicircle to provide the length of segment DE.

$$P \approx 9.42 + 9.42 + 6 + 6$$

$$P \approx 30.84$$

The length of the two smaller semicircles path is about 30.84 km.

Dan can bike ride on either path because they cover the same distance. Neither path is longer than the other one.

1. Find the circumference.

    a. Give an exact answer in terms of π.

    b. Use $\pi \approx \dfrac{22}{7}$ and express your answer as a fraction in lowest terms.

    c. Use *the* π button on your calculator, and express your answer to the nearest hundredth.

$$\frac{c}{28} = \frac{22}{7} \qquad \frac{7c}{7} = \frac{616}{7}$$

14 cm

2. Find the circumference.

    a. Give an exact answer in terms of π.

    b. Use $\pi \approx \dfrac{22}{7}$, and express your answer as a fraction in lowest terms.

$$\frac{c}{42} = \frac{22}{7} \qquad \frac{7c}{7} = \frac{924}{7}$$

132 cm

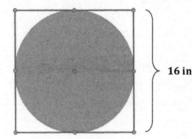

42 cm

3. The figure shows a circle within a square. Find the circumference of the circle. Let $\pi \approx 3.14$.

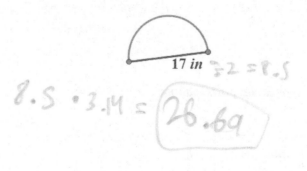

16 in

4. Consider the diagram of a semicircle shown.

    a. Explain in words how to determine the perimeter of a semicircle.

    use formla for c by not6

    b. Using d to represent the diameter of the circle, write an algebraic equation that will result in the perimeter of a semicircle.

    $\pi d \cdot \pi = c \div 2$

    c. Write another algebraic equation to represent the perimeter of a semicircle using r to represent the radius of a semicircle.

    $r \cdot \pi = \frac{1}{2}c$

    r

    d

5. Find the perimeter of the semicircle. Let $\pi \approx 3.14$.

    17 in $\div 2 = 8.5$

    $8.5 \cdot 3.14 = 26.69$

6. Ken's landscape gardening business makes odd-shaped lawns that include semicircles. Find the length of the edging material needed to border the two lawn designs. Use 3.14 for π.

    a. The radius of this flower bed is 2.5 m.

    b. The diameter of the semicircular section is 10 m, and the lengths of the sides of the two sides are 6 m.

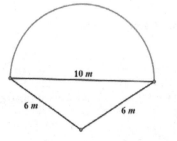

7. Mary and Margaret are looking at a map of a running path in a local park. Which is the shorter path from E to F, along the two semicircles or along the larger semicircle? If one path is shorter, how much shorter is it? Let $\pi \approx 3.14$.

8. Alex the electrician needs 34 yards of electrical wire to complete a job. He has a coil of wiring in his workshop. The coiled wire is 18 inches in diameter and is made up of 21 circles of wire. Will this coil be enough to complete the job? Let $\pi \approx 3.14$.

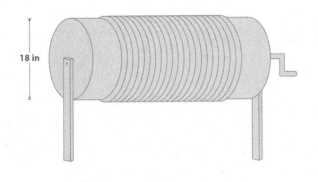

Exercises 1–3

Solve the problem below individually. Explain your solution.

1. Find the radius a circle if its circumference is 37.68 inches. Use $\pi \approx 3.14$.

2. Determine the area of the rectangle below. Name two ways that can be used to find the area of the rectangle.

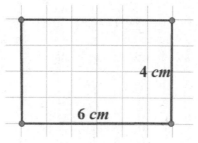

4 cm

6 cm

3. Find the length of a rectangle if the area is 27 cm^2 and the width is 3 cm.

Exploratory Challenge

To find the formula for the area of a circle, cut a circle into 16 equal pieces.

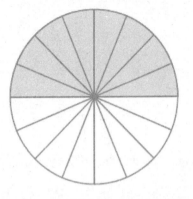

Arrange the triangular wedges by alternating the "triangle" directions and sliding them together to make a "parallelogram." Cut the triangle on the left side in half on the given line, and slide the outside half of the triangle to the other end of the parallelogram in order to create an approximate "rectangle."

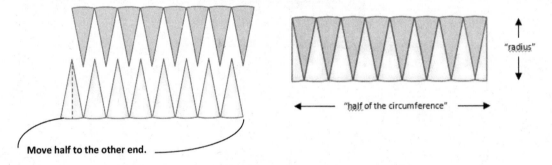

The circumference is $2\pi r$, where the radius is r. Therefore, half of the circumference is πr.

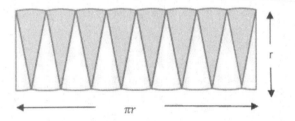

What is the area of the "rectangle" using the side lengths above?

Are the areas of the "rectangle" and the circle the same?

If the area of the rectangular shape and the circle are the same, what is the area of the circle?

Example 1

Use the shaded square centimeter units to approximate the area of the circle.

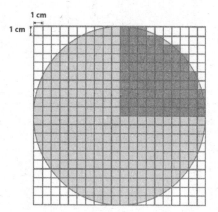

What is the radius of the circle?

What would be a quicker method for determining the area of the circle other than counting all of the squares in the entire circle?

Using the diagram, how many squares were used to cover one-fourth of the circle?

What is the area of the entire circle?

Example 2

A sprinkler rotates in a circular pattern and sprays water over a distance of 12 feet. What is the area of the circular region covered by the sprinkler? Express your answer to the nearest square foot.

Draw a diagram to assist you in solving the problem. What does the distance of 12 feet represent in this problem?

What information is needed to solve the problem?

Example 3

Suzanne is making a circular table out of a square piece of wood. The radius of the circle that she is cutting is 3 feet. How much waste will she have for this project? Express your answer to the nearest square foot.

Draw a diagram to assist you in solving the problem. What does the distance of 3 feet represent in this problem?

What information is needed to solve the problem?

Lesson 17:   The Area of a Circle

What information do we need to determine the area of the square and the circle?

How will we determine the waste?

Does your solution answer the problem as stated?

Exercises 4–6

4. A circle has a radius of 2 cm.

    a. Find the exact area of the circular region.

    b. Find the approximate area using 3.14 to approximate π.

5. A circle has a radius of 7 cm.

    a. Find the exact area of the circular region.

b. Find the approximate area using $\dfrac{22}{7}$ to approximate π.

c. What is the circumference of the circle?

6. Joan determined that the area of the circle below is 400π cm^2. Melinda says that Joan's solution is incorrect; she believes that the area is 100π cm^2. Who is correct and why?

20 cm

Relevant Vocabulary

CIRCULAR REGION (OR DISK): Given a point C in the plane and a number $r > 0$, the *circular region (or disk) with center C and radius r* is the set of all points in the plane whose distance from the point C is less than or equal to r.

The boundary of a disk is a circle. The *area of a circle* refers to the area of the disk defined by the circle.

EUREKA
MATH

Name _____   Date _____

Complete each statement using the words or algebraic expressions listed in the word bank below.

1. The length of the _____ of the rectangular region approximates the length of the _____ of the circle.

2. The _____ of the rectangle approximates the length of one-half of the circumference of the circle.

3. The circumference of the circle is _____.

4. The _____ of the _____ is $2r$.

5. The ratio of the circumference to the diameter is _____.

6. Area (circle) = Area of (_____) = $\frac{1}{2} \cdot$ circumference $\cdot r = \frac{1}{2}(2\pi r) \cdot r = \pi \cdot r \cdot r =$ _____.

Word bank				
radius	height	base	$2\pi r$	
diameter	circle	rectangle	πr^2	π

1. Find the area of the circle. Use $\frac{22}{7}$ as an approximation for π.

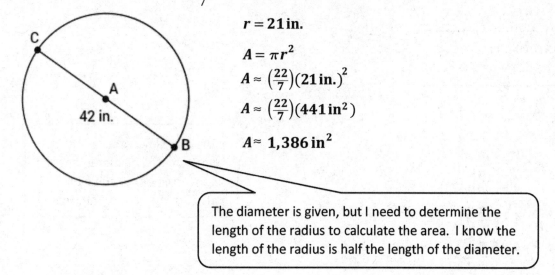

$$r = 21 \text{ in.}$$

$$A = \pi r^2$$

$$A \approx \left(\tfrac{22}{7}\right)(21 \text{ in.})^2$$

$$A \approx \left(\tfrac{22}{7}\right)(441 \text{ in}^2)$$

$$A \approx 1{,}386 \text{ in}^2$$

> The diameter is given, but I need to determine the length of the radius to calculate the area. I know the length of the radius is half the length of the diameter.

2. A circle has a diameter of 14 cm.

   a. Find the exact area, and find an approximate area using $\pi \approx 3.14$.

   $$r = 7 \text{ cm}$$

   Exact Area:

   $$A = \pi r^2$$

   $$A = \pi (7 \text{ cm})^2$$

   $$A = 49\pi \text{ cm}^2$$

   > π is an irrational number, so I must leave π in my answer to provide the exact area.

   Approximate Area:

   $$A \approx 49(3.14) \text{ cm}^2$$

   $$A \approx 153.86 \text{ cm}^2$$

   > Using the approximation of 3.14 for π provides an approximate area.

   b. What is the circumference of the circle using $\pi \approx 3.14$?

   $$d \approx 14 \text{ cm}$$

   $$C \approx (3.14)(14 \text{ cm})$$

   $$C \approx 43.96 \text{ cm}$$

   > I remember from the previous lesson that the formula for circumference is $C = \pi d$, which means I need to use the given diameter.

3. A circle has a circumference of 264 ft. Approximate the area of the circle, Use $\pi \approx \frac{22}{7}$.

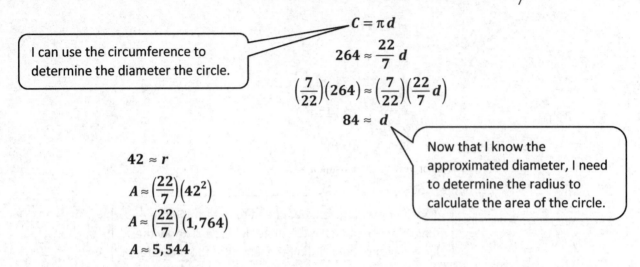

$$C = \pi d$$

$$264 \approx \frac{22}{7}d$$

$$\left(\frac{7}{22}\right)(264) \approx \left(\frac{7}{22}\right)\left(\frac{22}{7}d\right)$$

$$84 \approx d$$

I can use the circumference to determine the diameter the circle.

Now that I know the approximated diameter, I need to determine the radius to calculate the area of the circle.

$$42 \approx r$$

$$A \approx \left(\frac{22}{7}\right)\left(42^2\right)$$

$$A \approx \left(\frac{22}{7}\right)(1{,}764)$$

$$A \approx 5{,}544$$

The area of the circle is approximately $5{,}544$ ft^2.

4. The area of a circle is 81π in^2. Find its circumference.

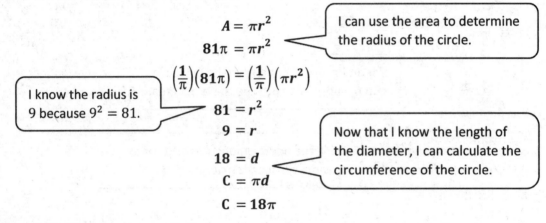

$$A = \pi r^2$$

$$81\pi = \pi r^2$$

$$\left(\frac{1}{\pi}\right)(81\pi) = \left(\frac{1}{\pi}\right)\left(\pi r^2\right)$$

$$81 = r^2$$

$$9 = r$$

$$18 = d$$

$$C = \pi d$$

$$C = 18\pi$$

I can use the area to determine the radius of the circle.

I know the radius is 9 because $9^2 = 81$.

Now that I know the length of the diameter, I can calculate the circumference of the circle.

The circumference of the circle is 18π in.

5. Find the ratio of the area of two circles with radii 5 in. and 6 in.

The area of the circle with radius 5 in. is 25π in^2. The area of the circle with radius 6 in. is 36π in^2. The ratio of the area of the two circles is 25π: 36π or 25: 36.

I calculate the area of each circle and then write the two areas as a ratio.

EUREKA MATH

1. The following circles are not drawn to scale. Find the area of each circle. (Use $\frac{22}{7}$ as an approximation for π.)

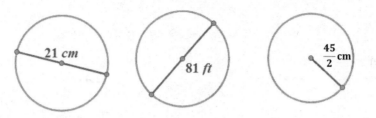

2. A circle has a diameter of 20 inches.
    a.   Find the exact area, and find an approximate area using $\pi \approx 3.14$.
    b.   What is the circumference of the circle using $\pi \approx 3.14$?

3. A circle has a diameter of 11 inches.
    a.   Find the exact area and an approximate area using $\pi \approx 3.14$.
    b.   What is the circumference of the circle using $\pi \approx 3.14$?

4. Using the figure below, find the area of the circle.

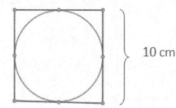

10 cm

5. A path bounds a circular lawn at a park. If the inner edge of the path is 132 ft. around, approximate the amount of area of the lawn inside the circular path. Use $\pi \approx \frac{22}{7}$.

6. The area of a circle is 36π cm^2. Find its circumference.

7. Find the ratio of the area of two circles with radii 3 cm and 4 cm.

8. If one circle has a diameter of 10 cm and a second circle has a diameter of 20 cm, what is the ratio of the area of the larger circle to the area of the smaller circle?

9. Describe a rectangle whose perimeter is 132 ft. and whose area is less than 1 ft^2. Is it possible to find a circle whose circumference is 132 ft. and whose area is less than 1 ft^2? If not, provide an example or write a sentence explaining why no such circle exists.

10. If the diameter of a circle is double the diameter of a second circle, what is the ratio of area of the first circle to the area of the second?

$$r = \frac{d}{2}$$

$$\pi r^2$$

$$\left[\pi \times \left(\frac{d}{2}\right)^2 \right] \frac{1}{2}$$

$$d = \frac{r}{2}$$

$$d^2 = \left(\frac{r}{2} \times \frac{1}{2}\right) \frac{1}{2} \qquad \pi \times \frac{d^2}{4} \times \frac{1}{2}$$

$$= \frac{r^2}{4} \times \frac{1}{2} = \boxed{\dfrac{\pi \, d^2}{8}}$$

$$
\begin{array}{r}
145.74 \\
\times \quad 6 \\
\hline
874.44
\end{array}
$$

$$
\begin{array}{r}
49.74 \\
\times \quad 6 \\
\hline
298.44
\end{array}
$$

Opening Exercise

Draw a circle with a diameter of 12 cm and a square with a side length of 12 cm on grid paper. Determine the area of the square and the circle.

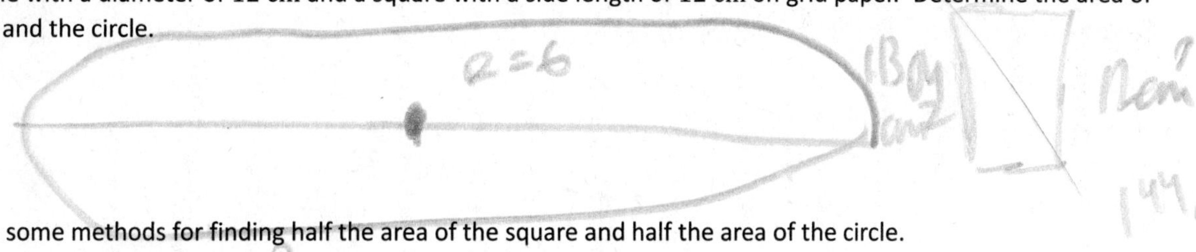

[handwritten: $r = 6$] *[handwritten: 113.04 cm², new², 144 cm²]*

Brainstorm some methods for finding half the area of the square and half the area of the circle.

[handwritten: $\pi r^2/2$ $\pi d/2$ πr]

Find the area of half of the square and half of the circle, and explain to a partner how you arrived at the area.

[handwritten: square 72   circle 56.52]

What is the ratio of the new area to the original area for the square and for the circle?

[handwritten: 1:2]

Find the area of one-fourth of the square and one-fourth of the circle, first by folding and then by another method. What is the ratio of the new area to the original area for the square and for the circle?

[handwritten: circle 28.26   square 36]

Write an algebraic expression that expresses the area of a semicircle and the area of a quarter circle.

[handwritten: $\frac{1}{2}a = \pi r^2/2$ $b = \frac{a}{4}$ or $\pi r^2/4$]

$\frac{7}{7} = 1$

$\frac{7}{7} = 1$

Example 1

Find the area of the following semicircle. Use $\pi \approx \frac{22}{7}$.

14 cm

$a = \frac{22}{7} 7^2 / 2$

$= \frac{22}{7} \times \frac{(7 \times 7)}{2}$

$\frac{7}{9} \left| \frac{\approx 22}{7} \right.$

$= 11 \times 7 = 77 \, cm^2$

What is the area of the quarter circle? Use $\pi \approx \frac{22}{7}$.

r = 6 cm

$\frac{22}{7} \times \frac{(6 \times 6)}{4} = \frac{22}{7} \times \frac{9}{7} = \frac{198}{7} = 28.3$

$a = 28.3 \, cm$

Example 2

Marjorie is designing a new set of placemats for her dining room table. She sketched a drawing of the placement on graph paper. The diagram represents the area of the placemat consisting of a rectangle and two semicircles at either end. Each square on the grid measures 4 inches in length.

Find the area of the entire placemat. Explain your thinking regarding the solution to this problem.

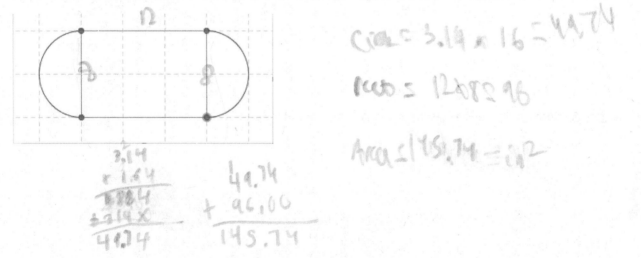

Circle $= 3.14 \times 16 = 49.74$

rect $= 12 \times 8 = 96$

Area $= 145.74 \, in^2$

```
  3.14
x 1.54
 .284
+314x
 4.94
```

```
  49.74
+ 96.00
 145.74
```

If Marjorie wants to make six placemats, how many square inches of fabric will she need? Assume there is no waste.

$145.74 \times 6 = 874.44 \, in^2$

Marjorie decides that she wants to sew on a contrasting band of material around the edge of the placemats. How much band material will Marjorie need?

Example 3

The circumference of a circle is 24π cm. What is the exact area of the circle?

Draw a diagram to assist you in solving the problem.

What information is needed to solve the problem?

Next, find the area.

Exercises

1. Find the area of a circle with a diameter of 42 cm. Use $\pi \approx \dfrac{22}{7}$.

2. The circumference of a circle is 9π cm.

     a. What is the diameter?

     b. What is the radius?

     c. What is the area?

3. If students only know the radius of a circle, what other measures could they determine? Explain how students would use the radius to find the other parts.

4. Find the area in the rectangle between the two quarter circles if $AF = 7$ ft, $FB = 9$ ft, and $HD = 7$ ft. Use $\pi \approx \frac{22}{7}$.
    Each quarter circle in the top-left and lower-right corners have the same radius.

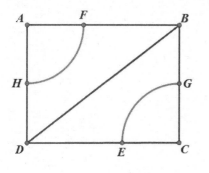

Name _____   Date _____

1. Ken's landscape gardening business creates odd-shaped lawns that include semicircles. Find the area of this semicircular section of the lawn in this design. Use $\frac{22}{7}$ for π.

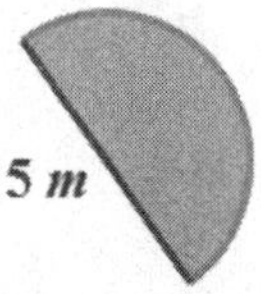

$$\frac{22}{7} \times \frac{(2.5 \times 2.5)}{2} = \frac{11}{7} \times \frac{6.25}{1}$$

$$5\,m$$

$$= \frac{68.75}{7} = 9.82$$

2. In the figure below, Ken's company has placed sprinkler heads at the center of the two small semicircles. The radius of the sprinklers is 5 ft. If the area in the larger semicircular area is the shape of the entire lawn, how much of the lawn will not be watered? Give your answer in terms of π and to the nearest tenth. Explain your thinking.

$$3.14 \times 25 / 2 = 39.25$$

$$39.25 - 25/2 =$$

I first got area of whole lawn than subtracted diameter of the 2 sprinklers combined to only get the non watered part

No place

1. Frederick is replacing a broken window that has a semicircle on top and a square on the bottom. He knows that the semicircular region has an area of 100.48 in^2.

   a. Draw a picture to represent the window.

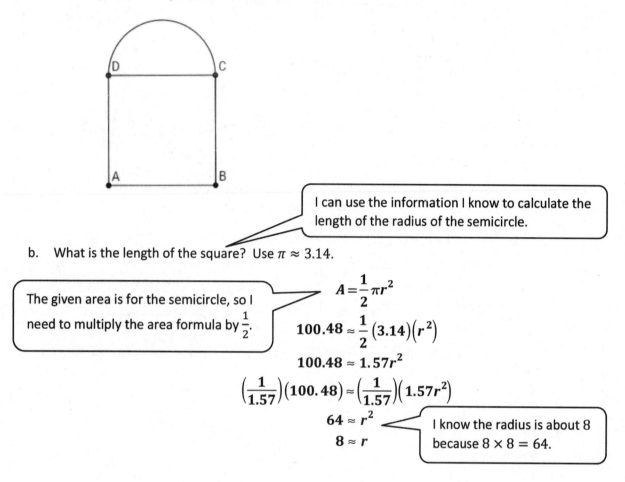

   I can use the information I know to calculate the length of the radius of the semicircle.

   b. What is the length of the square? Use $\pi \approx 3.14$.

   The given area is for the semicircle, so I need to multiply the area formula by $\frac{1}{2}$.

   $$A = \frac{1}{2}\pi r^2$$

   $$100.48 \approx \frac{1}{2}(3.14)(r^2)$$

   $$100.48 \approx 1.57 r^2$$

   $$\left(\frac{1}{1.57}\right)(100.48) \approx \left(\frac{1}{1.57}\right)(1.57 r^2)$$

   $$64 \approx r^2$$

   $$8 \approx r$$

   I know the radius is about 8 because $8 \times 8 = 64$.

   The length of the diameter is approximately 16 in., which means the length of the side of the square is approximately 16 in.

   The side length of the square is the same length as the diameter of the semicircle.

2. The diagram below is comprised of two squares and one quarter circle. It has a total length of 30 cm. What is the approximate area of the diagram? Use $\pi \approx 3.14$.

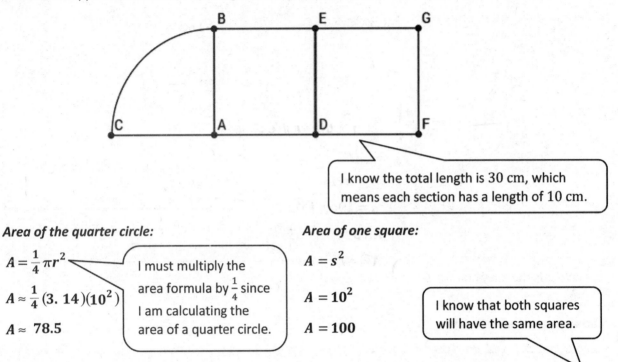

I know the total length is 30 cm, which means each section has a length of 10 cm.

Area of the quarter circle:

$A = \frac{1}{4}\pi r^2$

$A \approx \frac{1}{4}(3.14)(10^2)$

$A \approx 78.5$

I must multiply the area formula by $\frac{1}{4}$ since I am calculating the area of a quarter circle.

Area of one square:

$A = s^2$

$A = 10^2$

$A = 100$

I know that both squares will have the same area.

The total area of the diagram will be the sum of the area of the quarter circle and the area of the two squares.

$78.5 + 100 + 100 \approx 278.5$

Therefore, the sum of the diagram is approximately 278.5 cm^2.

3. The image below is the top of a unique end table. Help David determine the area of the table, so he can purchase a glass cover for the table. Approximate π as 3.14.

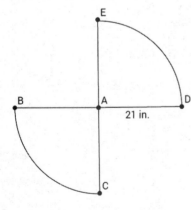

21 in.

Each part of the table represents a quarter circle, so I multiply the area formula by $\frac{1}{4}$.

$A = \frac{1}{4}\pi r^2$

$A \approx \frac{1}{4}(3.14)(21 \text{ in.})^2$

$A \approx 346.185 \text{ in}^2$

There are two quarter circles with the same area.

$A \approx 2\,(346.185 \text{ in}^2) \approx 692.37 \text{ in}^2$

The area of the entire table is approximately 692.37 in^2.

EUREKA MATH

4. Delecia is painting polk-a-dots in her daughter's room. Her daughter wants each one to have a purple center and a pink outline. Use the diagram below to determine the area of the pink paint. Use $\pi \approx 3.14$.

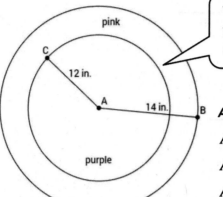

To find the area of pink paint, I need to find the area of the outside circle and the area of the inside circle.

Area of the outside circle:

$$A = \pi r^2$$
$$A = \pi (14 \text{ in.})^2$$
$$A = 196\,\pi \text{ in}^2$$

I use the exact area through all the calculations.

Area of the inside circle:

$$A = \pi r^2$$
$$A = \pi (12 \text{ in.})^2$$
$$A = 144\,\pi \text{ in}^2$$

To determine the area of just the pink paint, I need to find the difference of the two areas.

$$196\pi \text{ in}^2 - 144\pi \text{ in}^2 = 52\pi \text{ in}^2$$

The exact area of the pink circle is 52π in^2.

$$A \approx 52\,(3.14) \text{ in}^2 \approx 163.28 \text{ in}^2$$

The approximate area of the pink circle is 163.28 in^2.

$$a = \pi r^2 / 2$$

$$25.12 = \pi r^2 / 2$$

$$25.12 = 3.14 r^2 / 2$$

$$25.12 \times 2 = 3.14 \times \frac{r^2}{2} \times 2$$

$$50.24 = 3.14 \times r^2$$

$$\frac{50.24}{3.14} = \frac{3.14}{3.14} \times r^2$$

$$16 = r^2$$

$$4 = r^2$$

$$\boxed{r = 4} , \ +D = 8$$

1. Mark created a flower bed that is semicircular in shape, as shown in the image. The diameter of the flower bed is 5 m.

    a.   What is the perimeter of the flower bed? (Approximate π to be 3.14.)   7.85ₘ

    5 m

    b.   What is the area of the flower bed? (Approximate π to be 3.14.)

    9.8 m²

2. A landscape designer wants to include a semicircular patio at the end of a square sandbox. She knows that the area of the semicircular patio is 25.12 cm².

    a.   Draw a picture to represent this situation.

    b.   What is the length of the side of the square?

    8

3. A window manufacturer designed a set of windows for the top of a two-story wall. If the window is comprised of 2 squares and 2 quarter circles on each end, and if the length of the span of windows across the bottom is 12 feet, approximately how much glass will be needed to complete the set of windows?

    3.14 × 9/4

    = 7.06

    × 2

    14.12

    3   3

    3   3   3

    3   3   3   3

    12

    14.12

    + 18

    32.12

    14.12

4. Find the area of the shaded region. (Approximate π to be $\frac{22}{7}$.)

    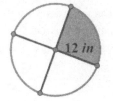

    12 in

    $\frac{22}{7} \times \frac{(12 \times 12)}{4} = \frac{22}{7} \times \frac{36}{1} = \frac{792}{7}$

    792 ÷ 7 = 113.14

5. The figure below shows a circle inside of a square. If the radius of the circle is 8 cm, find the following and explain your solution.

    a.   The circumference of the circle = 50.24 cm

    b.   The area of the circle = 200.96 cm²

    c.   The area of the square = 256 cm²   3.14 × 16

    3.14 × 64

    16 × 16

6. Michael wants to create a tile pattern out of three quarter circles for his kitchen backsplash. He will repeat the three quarter circles throughout the pattern. Find the area of the tile pattern that Michael will use. Approximate π as 3.14.

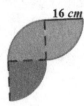

16 cm

7. A machine shop has a square metal plate with sides that measure 4 cm each. A machinist must cut four semicircles, with a radius of $\frac{1}{2}$ cm and four quarter circles with a radius of 1 cm from its sides and corners. What is the area of the plate formed? Use $\frac{22}{7}$ to approximate π.

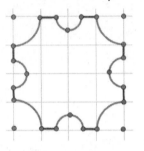

8. A graphic artist is designing a company logo with two concentric circles (two circles that share the same center but have different radii). The artist needs to know the area of the shaded band between the two concentric circles. Explain to the artist how he would go about finding the area of the shaded region.

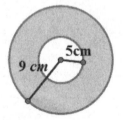

5cm
9 cm

9. Create your own shape made up of rectangles, squares, circles, or semicircles, and determine the area and perimeter.

Example: Area of a Parallelogram

The coordinate plane below contains figure P, parallelogram $ABCD$.

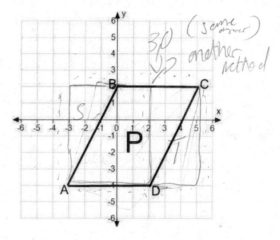

(handwritten, right margin) area of parallelogram
- turn it to a rectangle
- there will be two triangles to make a rectangle
- find area of rectangle and two triangles
- area of rectangle subtract with two triangles' areas

a.   Write the ordered pairs of each of the vertices next to the vertex points.

$A: (-3, -4)$ $B: (0, 2)$ $C: (5, 2)$

$D: (2, -4)$

b.   Draw a rectangle surrounding figure P that has vertex points of A and C. Label the two triangles in the figure as S and T.

c.   Find the area of the rectangle.

48 units^2

d.   Find the area of each triangle.

$S: 9$ units^2

$D: 9$ units^2

area(rect) - area(tri) = area(p)   P = parallelogram

e.   Use these areas to find the area of parallelogram $ABCD$.

30 units 2

The coordinate plane below contains figure R, a rectangle with the same base as the parallelogram above.

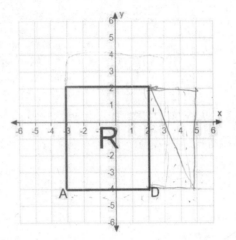

f.   Draw triangles S and T and connect to figure R so that you create a rectangle that is the same size as the rectangle you created on the first coordinate plane.

g.   Find the area of rectangle R.

48 30² units 2

h.   What do figures R and P have in common?

Same area

Exercises

1. Find the area of triangle ABC.

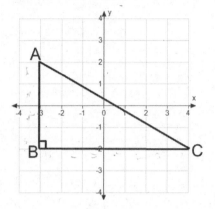

14 units²

2. Find the area of quadrilateral $ABCD$ two different ways.

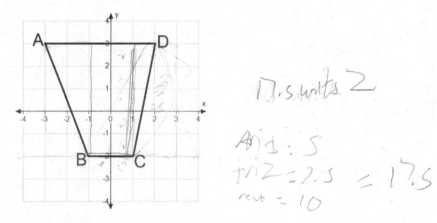

17.5 units²

Area: 5
Tri 2 = 2.5 = 17.5
rect = 10

3. The area of quadrilateral $ABCD$ is 12 sq. units. Find x.

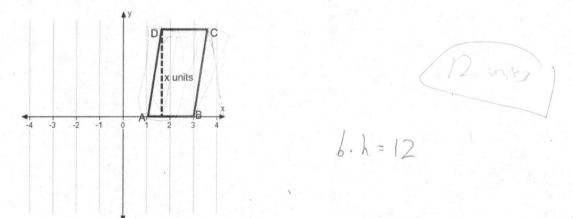

12 units

$b \cdot h = 12$

4. The area of triangle ABC is 14 sq. units. Find the length of side $\overline{BC}$.

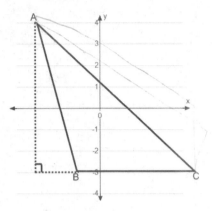

5. Find the area of triangle ABC.

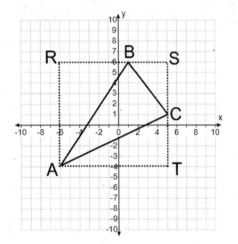

Lesson 19:   Unknown Area Problems on the Coordinate Plane

Name _____   Date _____

The figure *ABCD* is a rectangle. $AB = 2$ units, $AD = 4$ units, and $AE = FC = 1$ unit.

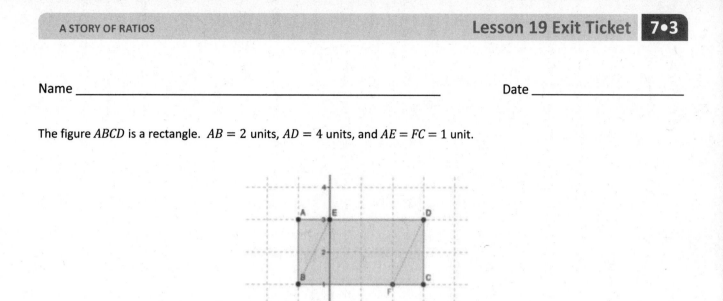

1. Find the area of rectangle *ABCD*.

2. Find the area of triangle *ABE*.   3. Find the area of triangle *DCF*.

4. Find the area of the parallelogram *BEDF* two different ways.

1. Find the area of each figure. When necessary, use 3.14 as an approximation for π.

a.

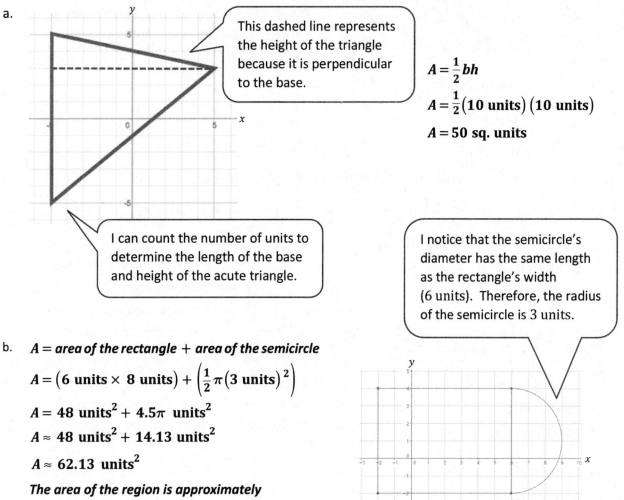

This dashed line represents the height of the triangle because it is perpendicular to the base.

$A = \frac{1}{2} bh$

$A = \frac{1}{2} (10 \text{ units}) (10 \text{ units})$

$A = 50 \text{ sq. units}$

I can count the number of units to determine the length of the base and height of the acute triangle.

I notice that the semicircle's diameter has the same length as the rectangle's width (6 units). Therefore, the radius of the semicircle is 3 units.

b. $A = $ *area of the rectangle $+$ area of the semicircle*

$A = \left(6 \text{ units} \times 8 \text{ units}\right) + \left(\frac{1}{2}\pi \left(3 \text{ units}\right)^2\right)$

$A = 48 \text{ units}^2 + 4.5\pi \text{ units}^2$

$A \approx 48 \text{ units}^2 + 14.13 \text{ units}^2$

$A \approx 62.13 \text{ units}^2$

The area of the region is approximately 62.13 units^2.

c.

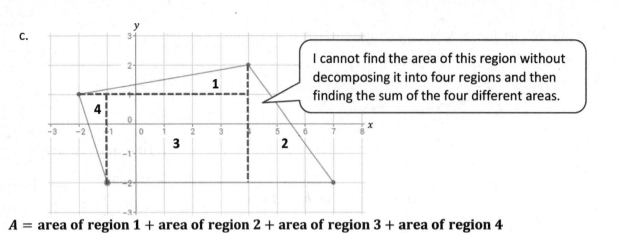

I cannot find the area of this region without decomposing it into four regions and then finding the sum of the four different areas.

A = area of region 1 + area of region 2 + area of region 3 + area of region 4

$$A = \left(\frac{1}{2} \times 1 \text{ unit} \times 6 \text{ units}\right) + \left(\frac{1}{2} \times 4 \text{ units} \times 3 \text{ units}\right) + \left(5 \text{ units} \times 3 \text{ units}\right) + \left(\frac{1}{2} \times 1 \text{ unit} \times 3 \text{ units}\right)$$

$$A = 3 \text{ units}^2 + 6 \text{ units}^2 + 15 \text{ units}^2 + 1.5 \text{ units}^2$$

$$A = 25.5 \text{ units}^2$$

To calculate the area of a triangle, I use the formula $A = \frac{1}{2}bh$. The area formula of rectangles is $A = lw$.

2. Draw a figure in the coordinate plane that matches the description.

A triangle with an area of 10 sq. units.

$$A = \frac{1}{2}bh$$

$$10 \text{ sq.units} = \frac{1}{2}bh$$

$$\frac{2}{1}(10 \text{ sq. units}) = \frac{2}{1}\left(\frac{1}{2}bh\right)$$

$$20 \text{ sq. units} = bh$$

One possible answer is a right triangle with a height of 5 units and a base of 4 units, which results in a product of 20 sq. units.

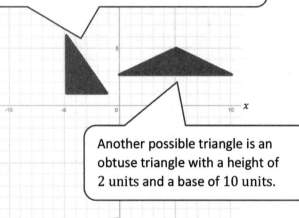

I can use my knowledge of solving equations to determine that the bh must equal 20 sq. units. This means that the base and the height must be factors of 20.

Another possible triangle is an obtuse triangle with a height of 2 units and a base of 10 units.

3. Find the area of triangle *DEF*.

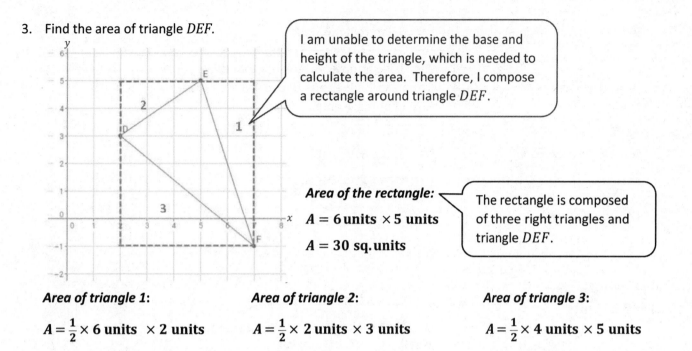

> I am unable to determine the base and height of the triangle, which is needed to calculate the area. Therefore, I compose a rectangle around triangle *DEF*.

Area of the rectangle:

$A = 6$ units $\times 5$ units

$A = 30$ sq. units

> The rectangle is composed of three right triangles and triangle *DEF*.

Area of triangle 1:

$A = \frac{1}{2} \times 6$ units $\times 2$ units

$A = 6$ sq. units

Area of triangle 2:

$A = \frac{1}{2} \times 2$ units $\times 3$ units

$A = 3$ sq. units

Area of triangle 3:

$A = \frac{1}{2} \times 4$ units $\times 5$ units

$A = 10$ sq. units

Area of triangle *DEF* = 30 sq. units − (6 sq. units + 3 sq. units + 10 sq. units)

Area of triangle *DEF* = 30 sq. units − 19 sq. units

Area of triangle *DEF* = 11 sq. units

> To determine the area of triangle *DEF*, I must subtract the areas of the three right triangles from the composed rectangle.

The area of triangle *DEF* is 11 sq. units.

4. Find the area of the quadrilateral using two different methods.

When I decompose, I break the shape into smaller shapes and then calculate the sum of these areas.

Method 1: Decompose

Area of the rectangle:

$A = 5 \text{ units} \times 4 \text{ units}$

$A = 20 \text{ sq. units}$

Area of the triangle:

$A = \frac{1}{2} \times 2 \text{ units} \times 4 \text{ units}$

$A = 4 \text{ sq. units}$

$A = 20 \text{ sq. units} + 4 \text{ sq. units}$

$A = 24 \text{ sq. units}$

I compose a larger shape that surrounds the original quadrilateral. I then calculate the difference of the area of the larger shape and the shape that is not included in the original quadrilateral.

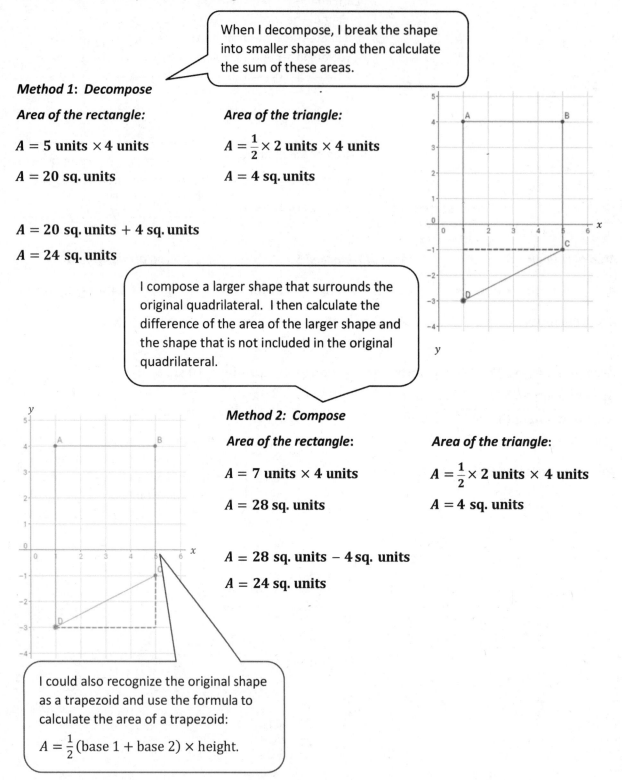

Method 2: Compose

Area of the rectangle:

$A = 7 \text{ units} \times 4 \text{ units}$

$A = 28 \text{ sq. units}$

Area of the triangle:

$A = \frac{1}{2} \times 2 \text{ units} \times 4 \text{ units}$

$A = 4 \text{ sq. units}$

$A = 28 \text{ sq. units} - 4 \text{ sq. units}$

$A = 24 \text{ sq. units}$

I could also recognize the original shape as a trapezoid and use the formula to calculate the area of a trapezoid:

$A = \frac{1}{2}(\text{base 1} + \text{base 2}) \times \text{height}.$

Find the area of each figure.

1.
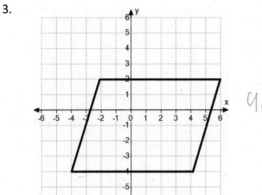

$27 \div 2 = 13 \cdot 5 \, units^2$

2.
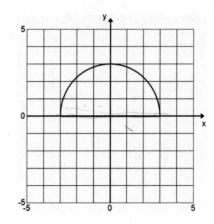

$3.14 \times 4.5 =$

3.
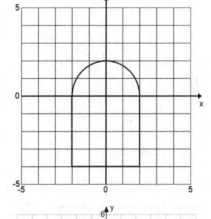

$9 \, units^2$

4.
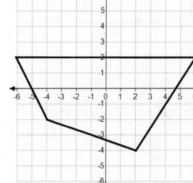

5.
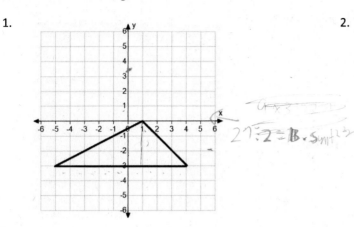

$12 + 56 =$
$68 \, units^2$

6.
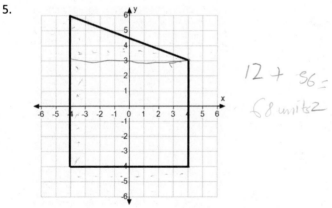

EUREKA
MATH®

For Problems 7–9, draw a figure in the coordinate plane that matches each description.

7. A rectangle with an area of 18 sq. units

8. A parallelogram with an area of 50 sq. units

9. A triangle with an area of 25 sq. units

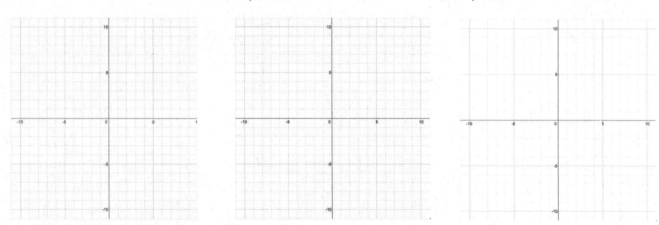

Find the unknown value labelled as *x* on each figure.

10. The rectangle has an area of 80 sq. units.

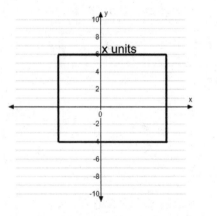

8

11. The trapezoid has an area of 115 sq. units.

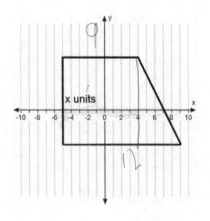

$(b_1 + b_2)/2 \times h$

$21/2 = 10.5 x$

$115 \div 10.5 = h = \boxed{10.95}$

12. Find the area of triangle *ABC*.

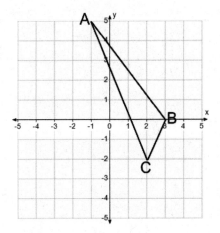

13. Find the area of the quadrilateral using two different methods. Describe the methods used, and explain why they result in the same area.

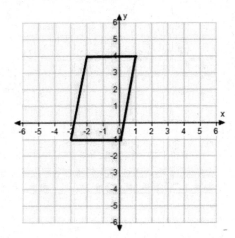

14. Find the area of the quadrilateral using two different methods. What are the advantages or disadvantages of each method?

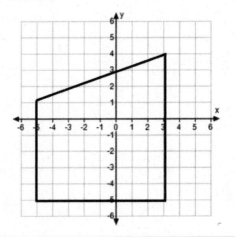

Example 1

Find the composite area of the shaded region. Use 3.14 for π.

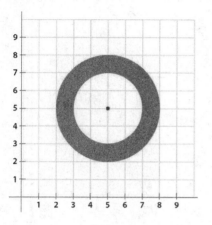

Exercise 1

A yard is shown with the shaded section indicating grassy areas and the unshaded sections indicating paved areas. Find the area of the space covered with grass in units^2.

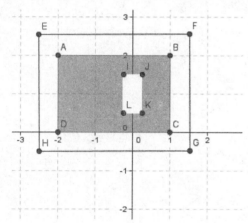

Example 2

Find the area of the figure that consists of a rectangle with a semicircle on top. Use 3.14 for π.

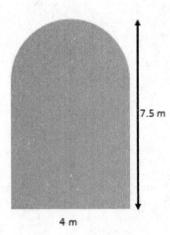

7.5 m

4 m

Exercise 2

Find the area of the shaded region. Use 3.14 for π.

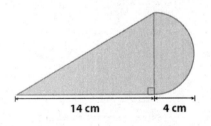

14 cm   4 cm

Lesson 20:   Composite Area Problems

Example 3

Find the area of the shaded region.

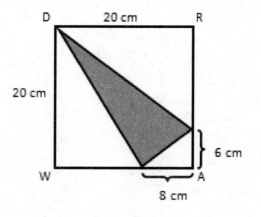

Redraw the figure separating the triangles; then, label the lengths discussing the calculations.

Exercise 3

Find the area of the shaded region. The figure is not drawn to scale.

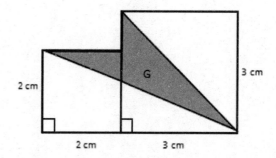

Name_____ Date _____

The unshaded regions are quarter circles. Approximate the area of the shaded region. Use $\pi \approx 3.14$.

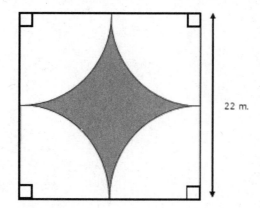

22 m.

1. The figure shows two semicircles. Find the area of the shaded region. Use 3.14 for π.

> I remember from previous lessons how to calculate the area of a circular region.

Area of the larger semicircle:

$A = \frac{1}{2}\pi r^2$

> The radius of the larger semicircle is 12 m.

$A \approx \frac{1}{2}(3.14)(12\text{ m})^2$

$A \approx 226.08\text{ m}^2$

12 m   12 m

Area of the smaller semicircle:

$A = \frac{1}{2}\pi r^2$

> The radius of the smaller semicircle is 6 m.

$A \approx \frac{1}{2}(3.14)(6\text{ m})^2$

$A \approx 56.52\text{ m}^2$

> The area of the shaded region is the difference between the area of each semicircle.

Area of the shaded region:

$A \approx 226.08\text{ m}^2 - 56.52\text{ m}^2$

$A \approx 169.56\text{ m}^2$

The approximate area of the shaded region is 169.56 m^2.

---

2. Find the area of the shaded region. Use 3.14 for π.

Area of the square:

$18 \text{ in.} \times 18 \text{ in.} = 324 \text{ in}^2$

Area of the semicircle:

$A = \frac{1}{2}\pi r^2$

$A \approx \frac{1}{2}(3.14)(9 \text{ in.})^2$

$A \approx 127.17 \text{ in}^2$

> In order to determine the area of the shaded region, I need to subtract the area of the semicircle from the area of the square.

> The diameter of the semicircle is the same length as a side of the square (18 in.), which means the radius of the semicircle is 9 in. because it is half of the diameter.

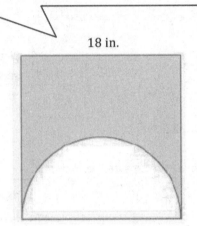

18 in.

Area of the shaded region:

$A \approx 324 \text{ in}^2 - 127.17 \text{ in}^2$

$A \approx 196.83 \text{ in}^2$

The area of the shaded region is approximately 196.83 in^2.

Lesson 20: Composite Area Problems

3. Sydney created a flower petal stencil to use to decorate the walls of her new daycare center. Sydney needs to calculate the area of each petal in order to plan the pattern on the wall. What is the area of Sydney's stencil. Provide your answer in terms of π.

Area of Region 1:

> The radius of Region 1 is half the diameter of 28 in.

$$A = \frac{1}{2}\pi r^2$$

$$A \approx \frac{1}{2}\pi(14\,\text{in.})^2$$

$$A \approx 98\pi\,\text{in}^2$$

> I decompose the petal into two different regions to make it easier to calculate the area.

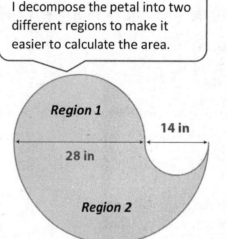

> Region 2 is the larger semicircle that has a smaller semicircle cut out of it. Therefore, I need to find the area of each semicircle and then calculate the difference.

Area of Region 2:

Let a represent the radius of the larger semicircle in region 2.
Let b represent the radius of the smaller semicircle in region 2.

> The diameter of the larger semicircle is 28 in. +14 in., or 42 in., which makes the radius 21 in.

$$A = \frac{1}{2}\pi a^2 - \frac{1}{2}\pi b^2$$

$$A = \frac{1}{2}\pi(21\,\text{in.})^2 - \frac{1}{2}\pi(7\,\text{in.})^2$$

$$A = 220.5\pi\,\text{in}^2 - 24.5\pi\,\text{in}^2$$

$$A = 196\pi\,\text{in}^2$$

Area of the flower petal:

$$A = 98\pi\,\text{in}^2 + 196\pi\,\text{in}^2$$

$$A = 294\pi\,\text{in}^2$$

> Now that I know the area of each region, I find the total area by calculating the sum of the areas of the two regions.

The exact area of the flower petal is $294\pi\,\text{in}^2$.

4. The figure is formed by five rectangles. Find the area of the unshaded rectangular region.

   area of the entire rectangle − area of the shaded rectangles = area of the unshaded rectangular region

   $A = 27\,\text{ft.} \times 17\,\text{ft.} - \left((27\,\text{ft.} \times 5\,\text{ft.}) + 2(12\,\text{ft.} \times 5\,\text{ft.}) + (12\,\text{ft.} \times 10\,\text{ft.})\right)$

   $A = 459\,\text{ft}^2 - \left(135\,\text{ft}^2 + 120\,\text{ft}^2 + 120\,\text{ft}^2\right)$

   $A = 459\,\text{ft}^2 - 375\,\text{ft}^2$

   $A = 84\,\text{ft}^2$

   The area of the unshaded rectangular region is 84 ft².

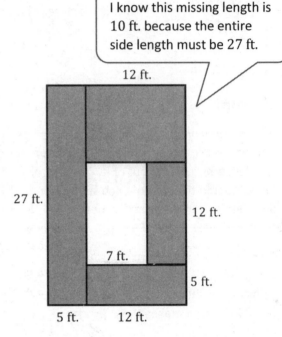

   I know this missing length is 10 ft. because the entire side length must be 27 ft.

   12 ft.

   27 ft.

   12 ft.

   7 ft.

   5 ft.

   5 ft.   12 ft.

5. The figure is a rectangle made out of triangles. Find the area of the shaded region.

   area of the shaded region = area of the rectangle − area of the unshaded triangles

   $A = 44\,\text{in.} \times 36\,\text{in.} - \left(\left(\frac{1}{2}\right)(18\,\text{in.} \times 44\,\text{in.}) + \left(\frac{1}{2}\right)(18\,\text{in.} \times 36\,\text{in.})\right)$

   $A = 1{,}584\,\text{in}^2 - \left(396\,\text{in}^2 + 324\,\text{in}^2\right)$

   $A = 1{,}584\,\text{in}^2 - 720\,\text{in}^2$

   $A = 864\,\text{in}^2$

   The area of the unshaded region is 864 in².

   18 in.

   44 in.

   18 in.

   36 in.

   The two unshaded regions are obtuse triangles, so I use the formula $A = \frac{1}{2}bh$ to calculate their areas.

1. Find the area of the shaded region. Use 3.14 for π.

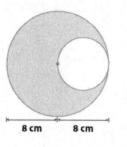

8 cm   8 cm

2. The figure shows two semicircles. Find the area of the shaded region. Use 3.14 for π.

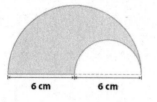

6 cm   6 cm

3. The figure shows a semicircle and a square. Find the area of the shaded region. Use 3.14 for π.

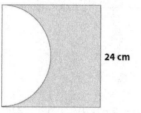

24 cm

4. The figure shows two semicircles and a quarter of a circle. Find the area of the shaded region. Use 3.14 for π.

10 cm   10 cm

5. Jillian is making a paper flower motif for an art project. The flower she is making has four petals; each petal is formed by three semicircles as shown below. What is the area of the paper flower? Provide your answer in terms of π.

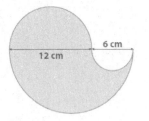

6. The figure is formed by five rectangles. Find the area of the unshaded rectangular region.

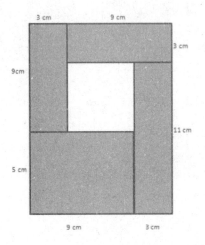

7. The smaller squares in the shaded region each have side lengths of 1.5 m. Find the area of the shaded region.

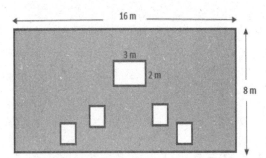

8. Find the area of the shaded region.

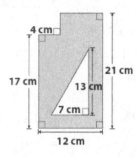

9.

a. Find the area of the shaded region.

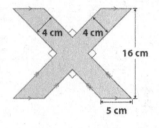

b. Draw two ways the figure above can be divided in four equal parts.

c. What is the area of one of the parts in (b)?

10. The figure is a rectangle made out of triangles. Find the area of the shaded region.

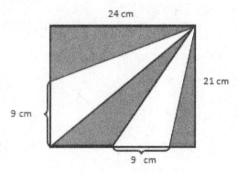

11. The figure consists of a right triangle and an eighth of a circle. Find the area of the shaded region. Use $\frac{22}{7}$ for π.

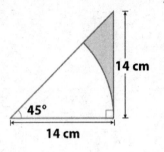

14 cm

45°

14 cm

Lesson 20:   Composite Area Problems

Opening Exercise: Surface Area of a Right Rectangular Prism

On the provided grid, draw a net representing the surfaces of the right rectangular prism (assume each grid line represents 1 inch). Then, find the surface area of the prism by finding the area of the net.

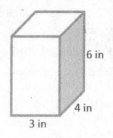

6 in

4 in

3 in

surface

area
is like the
perimeter
of a 3D
shape

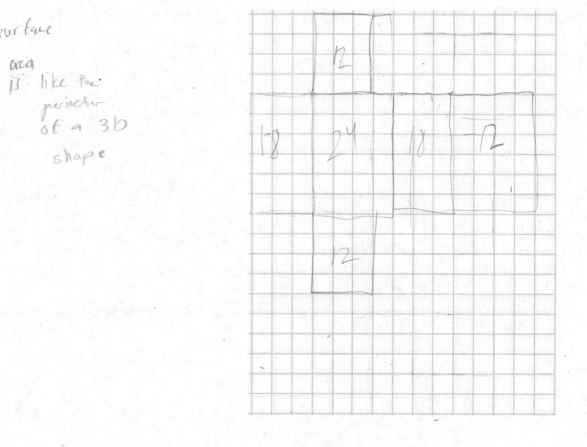

12

18   24   18   12

12

SA = 108 in.²

EUREKA MATH

© 2019 Great Minds®. eureka-math.org

SA: $2lw + 2lh + 2wh$ or $2(lw + lh + wh)$

Exercise 1

Marcus thinks that the surface area of the right triangular prism will be half that of the right rectangular prism and wants to use the modified formula $SA = \frac{1}{2}(2lw + 2lh + 2wh)$. Do you agree or disagree with Marcus? Use nets of the prisms to support your argument.

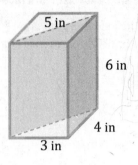
5 in
6 in
4 in
3 in

$06^2 \ 108 in^2$

$24 + 24 + 18 = 54 in^2$

Disagree formula does not include 5 in side length

Example 1: Lateral Area of a Right Prism

$SA = LA + 2B$

LA = Lateral Area
$2B = 2\times$ the area of base

A right triangular prism, a right rectangular prism, and a right pentagonal prism are pictured below, and all have equal heights of h.

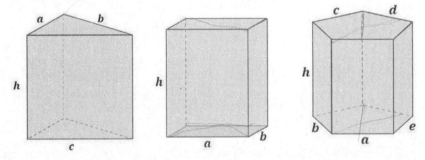

a. Write an expression that represents the lateral area of the right triangular prism as the sum of the areas of its lateral faces. *Lateral area is the surface area of a prism ignoring bases (top and bottom)*

$h \cdot c + a \cdot h + b \cdot h$
$ch + ah + bh$

$h(a + b + c)$

b.   Write an expression that represents the lateral area of the right rectangular prism as the sum of the areas of its lateral faces.

$$ah + bh + ah + bh$$

$$2(ah) + 2(bh)$$

c.   Write an expression that represents the lateral area of the right pentagonal prism as the sum of the areas of its lateral faces.

$$ah + bh + ch + dh + eh$$

$$h(a + b + c + d + e)$$

d.   What value appears often in each expression and why?

h, because it is the height of many faces

e.   Rewrite each expression in factored form using the distributive property and the height of each lateral face.

f.   What do the parentheses in each case represent with respect to the right prisms?

g.   How can we generalize the lateral area of a right prism into a formula that applies to all right prisms?

letters in Parentheses is perimeter of base
P = perimeter of base

$$LA: Ph$$ $$SA: Ph + 2b$$

P = a, b, c, d, e   (all same value
in before page)

Relevant Vocabulary

RIGHT PRISM: Let E and E' be two parallel planes. Let B be a triangular or rectangular region or a region that is the union of such regions in the plane E. At each point P of B, consider the segment PP' perpendicular to E, joining P to a point P' of the plane E'. The union of all these segments is a solid called a *right prism*.

There is a region B' in E' that is an exact copy of the region B. The regions B and B' are called the *base faces* (or just *bases*) of the prism. The rectangular regions between two corresponding sides of the bases are called *lateral faces* of the prism. In all, the boundary of a right rectangular prism has 6 *faces:* 2 base faces and 4 lateral faces. All adjacent faces intersect along segments called *edges* (base edges and lateral edges).

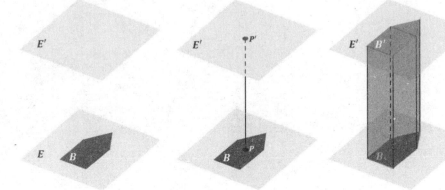

CUBE: A *cube* is a right rectangular prism all of whose edges are of equal length.

SURFACE: The *surface of a prism* is the union of all of its faces (the base faces and lateral faces).

NET: A *net* is a two-dimensional diagram of the surface of a prism.

1. Why are the lateral faces of right prisms always rectangular regions?

2. What is the name of the right prism whose bases are rectangles?

3. How does this definition of right prism include the interior of the prism?

Lesson 21:   Surface Area

Lesson Summary

The surface area of a right prism can be obtained by adding the areas of the lateral faces to the area of the bases. The formula for the surface area of a right prism is $SA = LA + 2B$, where SA represents the surface area of the prism, LA represents the area of the lateral faces, and B represents the area of one base. The lateral area LA can be obtained by multiplying the perimeter of the base of the prism times the height of the prism.

Name _____ Date _____

Find the surface area of the right trapezoidal prism. Show all necessary work.

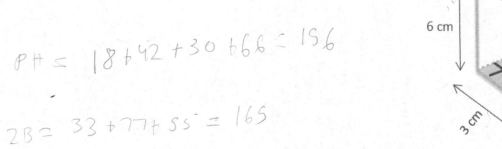

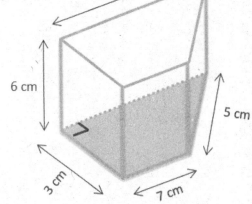

11 cm

6 cm

5 cm

3 cm

7 cm

$SA: LA + 2B$
$(PH$

$PH = 18 + 42 + 30 + 66 = 156$

$2B = 33 + 77 + 55 = 165$

$\begin{array}{r} 156 \\ +165 \\ \hline 321 \end{array}$

$SA = 321 \text{ cm}^2$

Surface Area of Nets

1. For the following net, draw a solid represented by the net, indicate the type of solid, and then find the solid's surface area.

Right triangular prism

$SA = LA + 2B$

> The figure has two identical bases, which means the net represents a prism. Since both of the bases are triangles, the figure is a triangular prism.

> In order to find the surface area, I find the area of the lateral faces and then add that to the area of both bases.

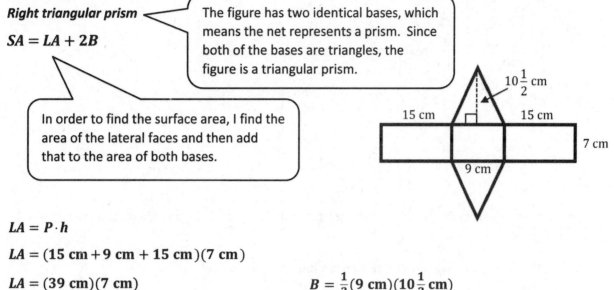

$LA = P \cdot h$

$LA = (15 \text{ cm} + 9 \text{ cm} + 15 \text{ cm})(7 \text{ cm})$

$LA = (39 \text{ cm})(7 \text{ cm})$

$LA = 273 \text{ cm}^2$

$B = \frac{1}{2}(9 \text{ cm})(10\frac{1}{2} \text{ cm})$

$B = 47.25 \text{ cm}^2$

$SA = 273 \text{ cm}^2 + 2(47.25 \text{ cm}^2)$

$SA = 273 \text{ cm}^2 + 94.5 \text{ cm}^2$

$SA = 367.5 \text{ cm}^2$

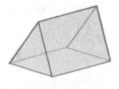

(3-Dimensional Form)

Surface Area of Cubes

2. Given the cube with edges that are $\frac{1}{2}$ inch long.

   a. Find the surface area of the cube.

   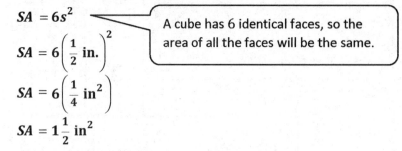

   $SA = 6s^2$

   A cube has 6 identical faces, so the area of all the faces will be the same.

   $SA = 6\left(\frac{1}{2} \text{ in.}\right)^2$

   $SA = 6\left(\frac{1}{4} \text{ in}^2\right)$

   $SA = 1\frac{1}{2} \text{ in}^2$

   b. Maria makes a scale drawing of the cube using a scale factor of 8. Find the surface area of the cube that Maria drew.

   $\frac{1}{2}$ in. $\cdot$ 8 = 4 in.; *the edge lengths of Maria's drawing would be 4 in.*

   I use the scale factor to determine the length of the edges in the scale drawing.

   $SA = 6(4 \text{ in.})^2$

   $SA = 6(16 \text{ in}^2)$

   $SA = 96 \text{ in}^2$

   c. What is the ratio of the surface area of the scale drawing to the surface area of the actual cube, and how does the value of the ratio compare to the scale factor?

   $96 \div 1\frac{1}{2}$

   I divide the two surface areas to determine the ratio.

   $96 \div \frac{3}{2}$

   $96 \times \frac{2}{3}$

   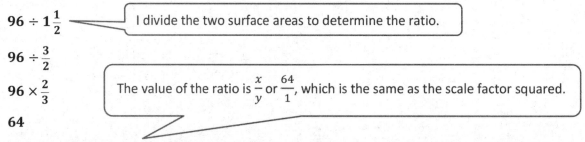

   The value of the ratio is $\frac{x}{y}$ or $\frac{64}{1}$, which is the same as the scale factor squared.

   64

   The ratio of the surface area of the scale drawing to the surface area of the actual cube is $64:1$. The value of the ratio is 64. The scale factor of the drawing is 8, and the value of the ratio of the surface area of the drawing to the surface area of the actual cube is 8^2, or 64.

Lesson 21:   Surface Area

EUREKA MATH

Surface Area of Prisms

3. Find the surface area of each of the following right prisms using the formula $SA = LA + 2B$.

   a. Trapezoidal Prism

   $$SA = LA + 2B$$

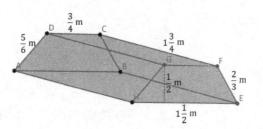

$$LA = \left(\frac{2}{3}m + \frac{3}{4}m + \frac{5}{6}m + 1\frac{1}{2}m\right)\left(1\frac{3}{4}m\right)$$

$$LA = \left(\frac{8}{12}m + \frac{9}{12}m + \frac{10}{12}m + 1\frac{6}{12}m\right)\left(1\frac{3}{4}m\right)$$

> I need common denominators before I can add fractions.

$$LA = \left(3\frac{3}{4}m\right)\left(1\frac{3}{4}m\right)$$

$$LA = \left(\frac{15}{4}m\right)\left(\frac{7}{4}m\right)$$

$$LA = \frac{105}{16}m^2$$

$$LA = 6\frac{9}{16}m^2$$

$$B = \frac{1}{2}(b_1 + b_2)h$$

$$B = \frac{1}{2}\left(\frac{3}{4}m + 1\frac{1}{2}m\right)\left(\frac{1}{2}m\right)$$

> The base of the prism is a trapezoid, so I use the area formula for a trapezoid to determine the area of each base.

$$B = \frac{1}{2}\left(2\frac{1}{4}m\right)\left(\frac{1}{2}m\right)$$

$$B = \frac{9}{16}m^2$$

$$SA = 6\frac{9}{16}m^2 + \frac{2}{1}\left(\frac{9}{16}m^2\right)$$

$$SA = 6\frac{9}{16}m^2 + \frac{18}{16}m^2$$

$$SA = 6\frac{9}{16}m^2 + 1\frac{2}{16}m^2$$

> When I multiply, I change the whole number to a fraction $\left(\frac{2}{1}\right)$ and then multiply the numerators and the denominators.

$$SA = 7\frac{11}{16}m^2$$

b. Kite Prism

$$SA = LA + 2B$$

> To find the area of the base, I decompose the base into two triangles.

$LA = (2.4\text{ in.} + 2.4\text{ in.} + 1.8\text{ in.} + 1.8\text{ in.})(4.2\text{ in.})$

$LA = (8.4\text{ in.})(4.2\text{ in.})$

$LA = 35.28\text{ in}^2$

$B = \frac{1}{2}(2.1\text{ in.} \times 1.1\text{ in.}) + \frac{1}{2}(1.6\text{ in.} \times 1.1\text{ in.})$

$B = \frac{1}{2}(2.31\text{ in}^2 + 1.76\text{ in}^2)$

$B = \frac{1}{2}(4.07\text{ in}^2)$

$B = 2.035\text{ in}^2$

> I need to use the given measurements to determine the height of the second triangle.
>
> 3.7 in. − 2.1 in. = 1.6 in.

$SA = 35.28\text{ in}^2 + 2(2.035\text{ in}^2)$

$SA = 35.28\text{ in}^2 + 4.07\text{ in}^2$

$SA = 39.35\text{ in}^2$

EUREKA
MATH

4. The surface area of the right rectangular prism is $164\frac{1}{2}$ ft^2. The dimensions of its base are 5 ft. and 8 ft. Use the formulas $SA = LA + 2B$ and $LA = Ph$ to find the unknown height, h, of the prism.

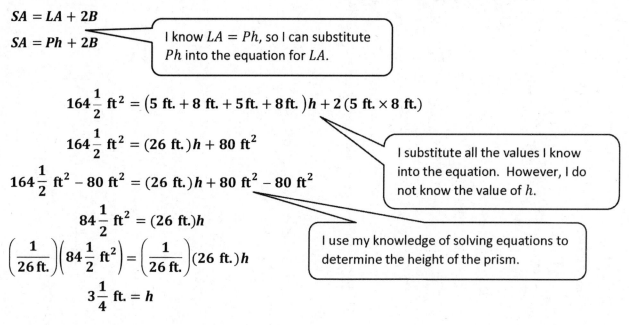

$$SA = LA + 2B$$
$$SA = Ph + 2B$$

> I know $LA = Ph$, so I can substitute Ph into the equation for LA.

$$164\frac{1}{2} \text{ ft}^2 = \left(5 \text{ ft.} + 8 \text{ ft.} + 5 \text{ ft.} + 8 \text{ ft.}\right)h + 2\left(5 \text{ ft.} \times 8 \text{ ft.}\right)$$

$$164\frac{1}{2} \text{ ft}^2 = (26 \text{ ft.})h + 80 \text{ ft}^2$$

> I substitute all the values I know into the equation. However, I do not know the value of h.

$$164\frac{1}{2} \text{ ft}^2 - 80 \text{ ft}^2 = (26 \text{ ft.})h + 80 \text{ ft}^2 - 80 \text{ ft}^2$$

$$84\frac{1}{2} \text{ ft}^2 = (26 \text{ ft.})h$$

$$\left(\frac{1}{26 \text{ ft.}}\right)\left(84\frac{1}{2} \text{ ft}^2\right) = \left(\frac{1}{26 \text{ ft.}}\right)(26 \text{ ft.})h$$

> I use my knowledge of solving equations to determine the height of the prism.

$$3\frac{1}{4} \text{ ft.} = h$$

The height of the prism is $3\frac{1}{4}$ ft.

© 2019 Great Minds®. eureka-math.org

1. For each of the following nets, highlight the perimeter of the lateral area, draw the solid represented by the net, indicate the type of solid, and then find the solid's surface area.

   a.

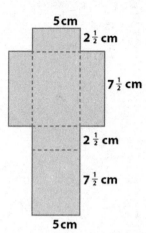

   b.

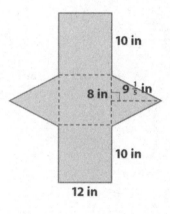

2. Given a cube with edges that are $\frac{3}{4}$ inch long:

   a. Find the surface area of the cube.

   b. Joshua makes a scale drawing of the cube using a scale factor of 4. Find the surface area of the cube that Joshua drew.

   c. What is the ratio of the surface area of the scale drawing to the surface area of the actual cube, and how does the value of the ratio compare to the scale factor?

3. Find the surface area of each of the following right prisms using the formula $SA = LA + 2B$.

a.

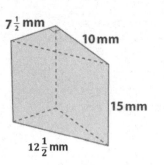

$7\frac{1}{2}$ mm

10 mm

15 mm

$12\frac{1}{2}$ mm

b.

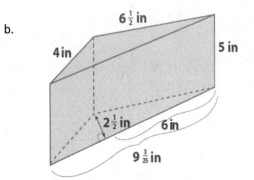

$6\frac{1}{2}$ in

4 in

5 in

$2\frac{1}{2}$ in

6 in

$9\frac{3}{25}$ in

c.

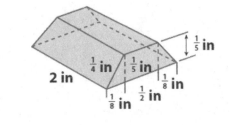

$\frac{1}{5}$ in

$\frac{1}{4}$ in

$\frac{1}{5}$ in

2 in

$\frac{1}{8}$ in

$\frac{1}{8}$ in

$\frac{1}{2}$ in

d.

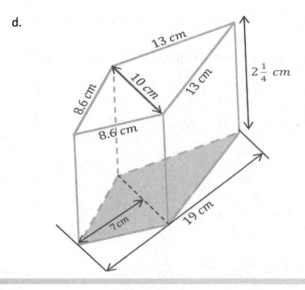

13 cm

10 cm

13 cm

$2\frac{1}{4}$ cm

8.6 cm

8.6 cm

7 cm

19 cm

Lesson 21: Surface Area

EUREKA MATH

4. A cube has a volume of 64 m^3. What is the cube's surface area?

5. The height of a right rectangular prism is $4\frac{1}{2}$ ft. The length and width of the prism's base are 2 ft. and $1\frac{1}{2}$ ft. Use the formula $SA = LA + 2B$ to find the surface area of the right rectangular prism.

6. The surface area of a right rectangular prism is $68\frac{2}{3} \text{ in}^2$. The dimensions of its base are 3 in. and 7 in. Use the formula $SA = LA + 2B$ and $LA = Ph$ to find the unknown height h of the prism.

7. A given right triangular prism has an equilateral triangular base. The height of that equilateral triangle is approximately 7.1 cm. The distance between the bases is 9 cm. The surface area of the prism is $319\frac{1}{2} \text{ cm}^2$. Find the approximate lengths of the sides of the base.

Opening Exercise

What is the area of the composite figure in the diagram? Is the diagram a net for a three-dimensional image? If so, sketch the image. If not, explain why.

Example 1

The pyramid in the picture has a square base, and its lateral faces are triangles that are exact copies of one another. Find the surface area of the pyramid.

Example 2: Using Cubes

There are 13 cubes glued together forming the solid in the diagram. The edges of each cube are $\frac{1}{4}$ inch in length. Find the surface area of the solid.

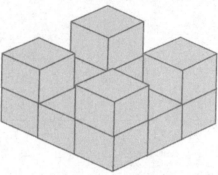

Example 3

Find the total surface area of the wooden jewelry box. The sides and bottom of the box are all $\frac{1}{4}$ inch thick.

What are the faces that make up this box?

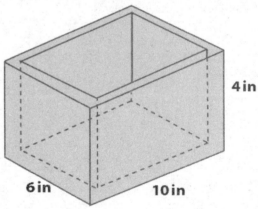

How does this box compare to other objects that you have found the surface area of?

Large Prism   *Small Prism*

Surface Area of the Box

Name _____   Date _____

1. The right hexagonal pyramid has a hexagon base with equal-length sides. The
    lateral faces of the pyramid are all triangles (that are exact copies of one
    another) with heights of 15 ft. Find the surface area of the pyramid.

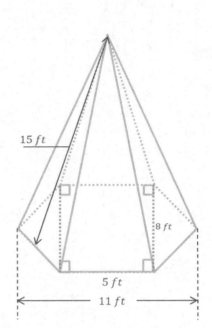

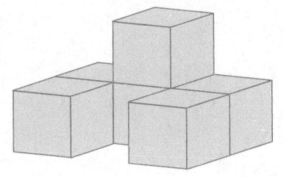

2. Six cubes are glued together to form the solid shown in the
    diagram. If the edges of each cube measure $1\frac{1}{2}$ inches in length,
    what is the surface area of the solid?

Surface Area of Nets

1. For the following net, draw (or describe) the solid represented by the net, and find its surface area. Each of the triangular faces is identical.

    The net represents a square pyramid where the four lateral faces are identical triangles. The base is square.

    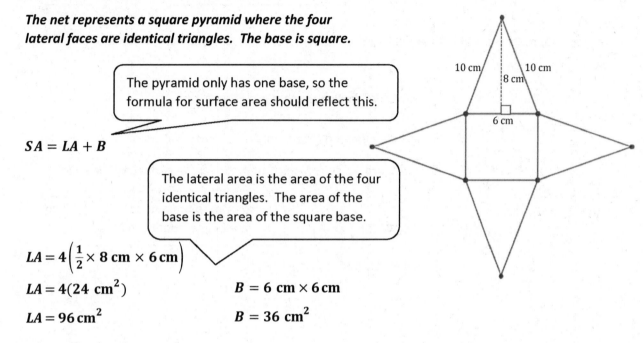

    The pyramid only has one base, so the formula for surface area should reflect this.

    $$SA = LA + B$$

    The lateral area is the area of the four identical triangles. The area of the base is the area of the square base.

    $$LA = 4\left(\frac{1}{2} \times 8 \text{ cm} \times 6 \text{ cm}\right)$$

    $$LA = 4(24 \text{ cm}^2) \qquad\qquad B = 6 \text{ cm} \times 6 \text{ cm}$$

    $$LA = 96 \text{ cm}^2 \qquad\qquad\quad B = 36 \text{ cm}^2$$

    $$SA = 96 \text{ cm}^2 + 36 \text{ cm}^2$$

    $$SA = 132 \text{ cm}^2$$

Surface Area of Multiple Cubes

I know the volume of a cube is s^3, which means each edge length is $\frac{1}{3}$ cm because $\left(\frac{1}{3}\text{ cm}\right)^3 = \frac{1}{27}\text{ cm}^3$.

2. In the diagram, there are 14 cubes glued together to form a solid. Each cube has a volume of $\frac{1}{27}$ cm^3. Find the surface area of the solid.

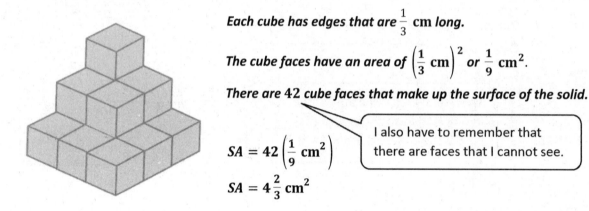

Each cube has edges that are $\frac{1}{3}$ cm long.

The cube faces have an area of $\left(\frac{1}{3}\text{ cm}\right)^2$ or $\frac{1}{9}$ cm^2.

There are 42 cube faces that make up the surface of the solid.

$$SA = 42\left(\frac{1}{9}\text{ cm}^2\right)$$

$$SA = 4\frac{2}{3}\text{ cm}^2$$

I also have to remember that there are faces that I cannot see.

Surface Area of Prisms with Shapes Cut Out

3. Find the surface area of the solid shown in the diagram. The solid is a right triangular prism (with right triangular bases) with a smaller right triangular prism removed from it.

$$SA = LA + 2B$$

$$LA = Ph$$

$$LA = \left(7\text{ m} + 7\text{ m} + 9\frac{9}{10}\text{ m}\right)(3\text{ m})$$

$$LA = \left(23\frac{9}{10}\text{ m}\right)(3\text{ m})$$

$$LA = 71\frac{7}{10}\text{ m}^2$$

The $7\frac{7}{9}$ m by $1\frac{1}{4}$ m rectangle has to be taken away from the lateral area.

$$A = 7\frac{7}{9}\text{ m} \times 1\frac{1}{4}\text{ m} \qquad LA = 71\frac{7}{10}\text{ m}^2 - 9\frac{13}{18}\text{ m}^2$$

$$A = 9\frac{13}{18}\text{ m}^2 \qquad\qquad LA = 61\frac{44}{45}\text{ m}^2$$

7 m
$5\frac{1}{2}$ m
$9\frac{9}{10}$ m
$7\frac{7}{9}$ m
$5\frac{1}{2}$ m
$1\frac{1}{4}$ m
7 m
3 m

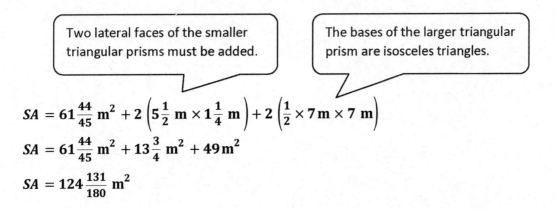

Two lateral faces of the smaller triangular prisms must be added.

The bases of the larger triangular prism are isosceles triangles.

$$SA = 61\frac{44}{45}\ \text{m}^2 + 2\left(5\frac{1}{2}\ \text{m} \times 1\frac{1}{4}\ \text{m}\right) + 2\left(\frac{1}{2} \times 7\ \text{m} \times 7\ \text{m}\right)$$

$$SA = 61\frac{44}{45}\ \text{m}^2 + 13\frac{3}{4}\ \text{m}^2 + 49\ \text{m}^2$$

$$SA = 124\frac{131}{180}\ \text{m}^2$$

4. The diagram below shows a cube that has had three square holes punched completely through the cube on three perpendicular axes. Find the surface area of the remaining solid.

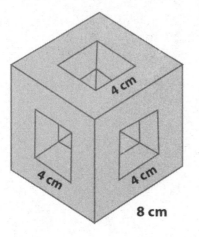

The exterior is a cube with a square hole on each of its six faces.

Surface area of the exterior:

$$SA = 6(8\ \text{cm} \times 8\ \text{cm}) - 6(4\ \text{cm} \times 4\ \text{cm})$$

$$SA = 6(64\ \text{cm}^2) - 6(16\ \text{cm}^2)$$

$$SA = 384\ \text{cm}^2 - 96\ \text{cm}^2$$

$$SA = 288\ \text{cm}^2$$

Surface area of the holes:

Each of the holes is a square with a depth of 2 cm.

$$SA = 6(LA)$$

$$SA = 6((4\ \text{cm} + 4\ \text{cm} + 4\ \text{cm} + 4\ \text{cm}) \times 2\ \text{cm})$$

$$SA = 6(16\ \text{cm} \times 2\ \text{cm})$$

$$SA = 6(32\ \text{cm}^2)$$

$$SA = 192\ \text{cm}^2$$

The total surface area would be the sum of the exterior surface area and the area of the holes.

Total surface area:

$$SA = 288\ \text{cm}^2 + 192\ \text{cm}^2$$

$$SA = 480\ \text{cm}^2$$

1. For each of the following nets, draw (or describe) the solid represented by the net and find its surface area.

   a. The equilateral triangles are exact copies.

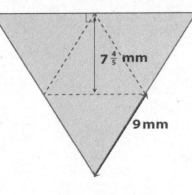

   b.

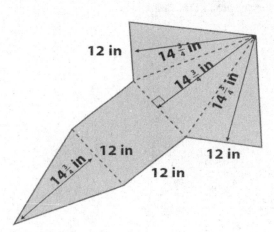

2. Find the surface area of the following prism.

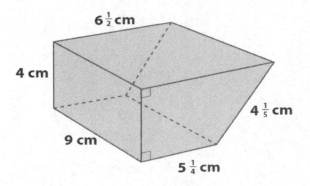

3. The net below is for a specific object. The measurements shown are in meters. Sketch (or describe) the object, and then find its surface area.

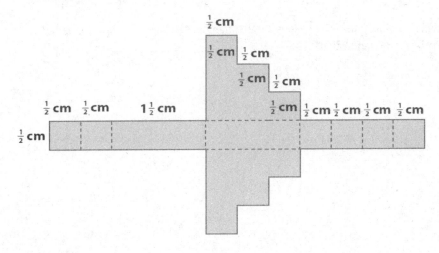

4. In the diagram, there are 14 cubes glued together to form a solid. Each cube has a volume of $\frac{1}{8}$ in³. Find the surface area of the solid.

5. The nets below represent three solids. Sketch (or describe) each solid, and find its surface area.

a.

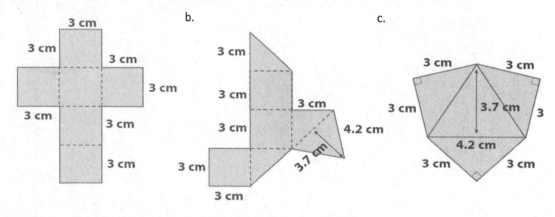

3 cm
3 cm   3 cm
3 cm
3 cm   3 cm
3 cm

b.

3 cm
3 cm
3 cm   3 cm
3 cm   4.2 cm
3 cm   3.7 cm
3 cm

c.

3 cm   3 cm
3 cm   3.7 cm   3
4.2 cm
3 cm   3 cm

d. How are figures (b) and (c) related to figure (a)?

6. Find the surface area of the solid shown in the diagram. The solid is a right triangular prism (with right triangular bases) with a smaller right triangular prism removed from it.

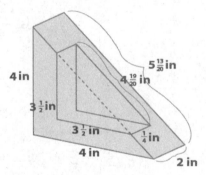

4 in
$5\frac{13}{20}$ in
$4\frac{19}{20}$ in
$3\frac{1}{2}$ in
$3\frac{1}{2}$ in $\frac{1}{4}$ in
4 in
2 in

7. The diagram shows a cubic meter that has had three square holes punched completely through the cube on three
    perpendicular axes. Find the surface area of the remaining solid.

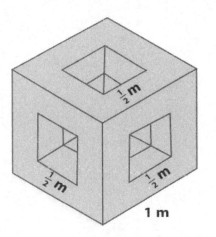

Opening Exercise

The volume of a solid is a quantity given by the number of unit cubes needed to fill the solid. Most solids—rocks, baseballs, people—cannot be filled with unit cubes or assembled from cubes. Yet such solids still have volume. Fortunately, we do not need to assemble solids from unit cubes in order to calculate their volume. One of the first interesting examples of a solid that cannot be assembled from cubes, but whose volume can still be calculated from a formula, is a right triangular prism.

What is the area of the square pictured on the right? Explain.

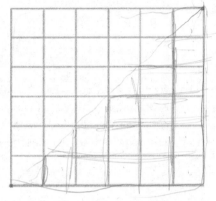

36 units^2

Draw the diagonal joining the two given points; then, darken the grid lines within the lower triangular region. What is the area of that triangular region? Explain.

$6 \times 6 \div \frac{1}{2} = \boxed{18}$

This triangle is half of 6 square (36/2 =18)

Exploratory Challenge: The Volume of a Right Prism   Volume = lengthx (reub prism

What is the volume of the right prism pictured on the right? Explain.

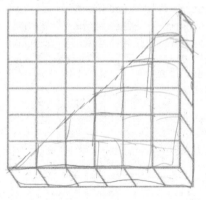

Volume = area of a 3D shape

$(0.6 \times) = 36$ units^3

Tot Volume = 18 units^3

Draw the same diagonal on the square base as done above; then, darken the grid lines on the lower right triangular prism. What is the volume of that right triangular prism? Explain.

18. units³ → half of 06

How could we create a right triangular prism with five times the volume of the right triangular prism pictured to the right, without changing the base? Draw your solution on the diagram, give the volume of the solid, and explain why your solution has five times the volume of the triangular prism.

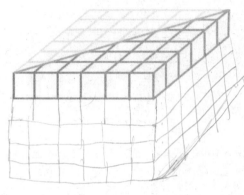

$6 \times 6 \times 5 = 180$

$180 \div 2 = \boxed{90}$

What could we do to cut the volume of the right triangular prism pictured on the right in half without changing the base? Draw your solution on the diagram, give the volume of the solid, and explain why your solution has half the volume of the given triangular prism.

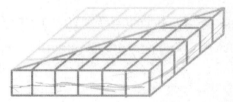

obvious

$\frac{1}{2} \times \frac{1}{2} = \frac{1}{4}$

$6 \times 6 \times \frac{1}{4} = \boxed{9}$

To find the volume (V) of any right prism ...

$V = Bh$ → height

area of base

© 2019 Great Minds®. eureka-math.org

Height connects bases together

Example: The Volume of a Right Triangular Prism

Find the volume of the right triangular prism shown in the diagram using $V = Bh$.

b base

triangle are the bases

$\frac{1}{2} = 3.25$

$B = 4 \lor \frac{1}{4} s \; |$

$V s \; 1 \cdot 6\frac{1}{2} = 6\frac{1}{2} \; m^3$

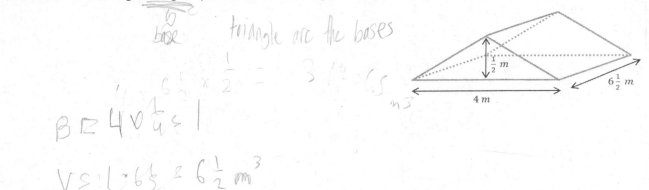

$\frac{1}{2}$ m

4 m

$6\frac{1}{2}$ m

m^3

Exercise: Multiple Volume Representations

The right pentagonal prism is composed of a right rectangular prism joined with a right triangular prism. Find the volume of the right pentagonal prism shown in the diagram using two different strategies.

Tri: $3.25 m^3$

$1 \times 6.5 s = 6\frac{1}{2} = 3.25$

Rect: $4 \times 6.5 \times 6.5 = 169 m^3$

42.25

$\times \quad 4$

$\overline{169.00}$

6.5

$\times 6.5$

$\overline{32 \; 5}$

$39 \; 0$

$\overline{42.25}$

$\boxed{V \leq 172.25 m^3}$

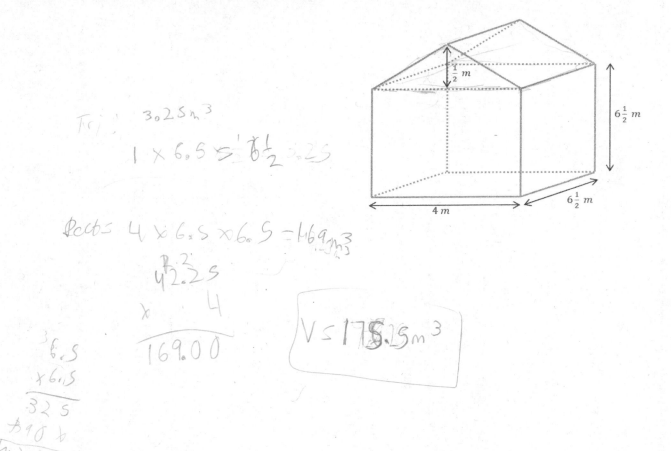

$\frac{1}{2}$ m

$6\frac{1}{2}$ m

4 m

$6\frac{1}{2}$ m

Name _____   Date _____

The base of the right prism is a hexagon composed of a rectangle and two triangles. Find the volume of the right hexagonal prism using the formula $V = Bh$.

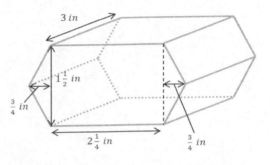

1. Calculate the volume of each right prism using the formula $V = Bh$.

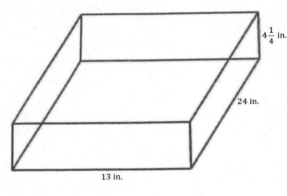

a.

> To calculate the volume of the prism, I find the product of the area of the base and the height of the prism.

$V = Bh$

$V = (13 \text{ in.} \times 24 \text{ in.}) \times 4\frac{1}{4} \text{ in.}$

$V = 312 \text{ in.}^2 \times 4\frac{1}{4} \text{ in.}$

$V = 1,326 \text{ in}^3$

The volume of the solid is $1,326$ in^3.

b. $B = A_{\text{lg rectangle}} - A_{\text{sm rectangle}}$

$B = \left(8\frac{1}{3} \text{ cm} \times 5 \text{ cm}\right) - (3 \text{ cm} \times 2 \text{ cm})$

$B = \left(\frac{25}{3} \text{ cm} \times 5 \text{ cm}\right) - 6 \text{ cm}^2$

$B = 41\frac{2}{3} \text{ cm}^2 - 6 \text{ cm}^2$

$B = 35\frac{2}{3} \text{ cm}^2$

$8\frac{1}{3}$ cm

$1\frac{1}{3}$ cm

2 cm 4 cm 5 cm

$1\frac{1}{3}$ cm

> There is a section cut out of the prism. I can find the area of the entire base and then subtract the area that has been cut out of the prism.

$V = Bh$

$V = 35\frac{2}{3} \text{ cm}^2\left(1\frac{1}{3} \text{ cm}\right)$

$V = \frac{107}{3} \text{ cm}^2\left(\frac{4}{3} \text{ cm}\right)$

$V = \frac{428}{9} \text{ cm}^3$

$V = 47\frac{5}{9} \text{ cm}^3$

> Once I calculate the area of the base, I still multiply it by the height of the prism to calculate the volume of the entire prism.

The volume of the solid is $47\frac{5}{9}$ cm^3.

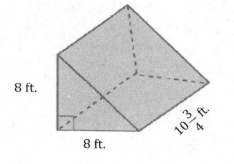

c.

$$B = \frac{1}{2}bh_{\text{triangle}}$$

$$B = \frac{1}{2}(8 \text{ ft.})(8 \text{ ft.})$$

$$B = 32 \text{ ft}^2$$

> The base of the prism is a triangle, so I use the area formula for a triangle to calculate the area of the base.

8 ft.

$10\frac{3}{4}$ ft.

8 ft.

$$V = Bh_{\text{prism}}$$

$$V = 32 \text{ ft}^2 \times 10\frac{3}{4} \text{ ft.}$$

$$V = 32 \text{ ft}^2 \times \frac{43}{4} \text{ ft.}$$

$$V = 344 \text{ ft}^3$$

> The height of the prism is the distance between the two triangular bases.

The volume of the solid is 344 ft³.

> I can decompose the trapezoidal base into a rectangle and a triangle. I could also choose to use the area formula for a trapezoid, which is $A = \frac{1}{2}(b_1 + b_2)h$.

d.

$15\frac{1}{2}$ in.

13 in.

18 in.

$13\frac{1}{5}$ in.

$14\frac{1}{4}$ in.

$$B = A_{\text{rectangle}} + A_{\text{triangle}}$$

$$B = \left(14\frac{1}{4} \text{ in.} \times 13 \text{ in.}\right) + \left(\frac{1}{2} \times 1\frac{1}{4} \text{ in.} \times 13 \text{ in.}\right)$$

$$B = \left(\frac{57}{4} \text{ in.} \times 13 \text{ in.}\right) + \left(\frac{1}{2} \times \frac{5}{4} \text{ in.} \times 13 \text{ in.}\right)$$

$$B = \frac{741}{4} \text{ in}^2 + \frac{65}{8} \text{ in}^2$$

$$B = \frac{1482}{8} \text{ in}^2 + \frac{65}{8} \text{ in}^2$$

$$B = \frac{1547}{8} \text{ in}^2$$

$$V = Bh$$

$$V = \frac{1547}{8} \text{ in}^2 \times 18 \text{ in.}$$

$$V = 3,480\frac{3}{4} \text{ in}^3$$

> I wait to convert this fraction into a mixed number until I am done with all of my calculations.

The volume of this solid is $3,480\frac{3}{4}$ in³.

2. Let l represent the length, w the width, and h the height of a right rectangular prism. Find the volume of the prism when $l = 12$ m, $w = \frac{5}{6}$ m, $h = 6\frac{1}{10}$ m.

$V = Bh$

$V = lwh$

$V = 12 \text{ m} \times \frac{5}{6}\text{m} \times 6\frac{1}{10}\text{ m}$

> I multiply the length and width to calculate the area of the base of a right rectangular prism. Therefore, I can substitute lw in for B in the volume formula.

$V = 10 \text{ m}^2 \times 6\frac{1}{10}\text{ m}$

$V = 61 \text{ m}^3$

3. Find the length of the edge indicated in the diagram if the volume of the prism is 432 cm³.

   Let h represent the number of centimeters in the height of the triangular base of the prism.

9 cm

?

12 cm

$$V = Bh$$

$$V = \left(\frac{1}{2}bh_{\text{triangle}}\right)(h_{\text{prism}})$$

$$432\,\text{cm}^3 = \left(\frac{1}{2} \cdot 9\,\text{cm} \cdot h\right)(12\,\text{cm})$$

$$432\,\text{cm}^3 = \frac{1}{2} \cdot 108\,\text{cm}^2 \cdot h$$

$$432\,\text{cm}^3 = 54\,\text{cm}^2 \cdot h$$

$$\left(\frac{1}{54\,\text{cm}^2}\right)(432\,\text{cm}^3) = \left(\frac{1}{54\,\text{cm}^2}\right)54\,\text{cm}^2 \cdot h$$

$$8\,\text{cm} = h$$

> I substitute the information I know into the volume formula and then use my knowledge of solving equations to determine the height of the prism.

The height of the triangle is 8 cm.

4. Given a right rectangular prism with a volume of 36 in³, a length of 8 in., and a width of 3 in., find the height of the prism.

    Let h represent the number of inches in the height of the prism.

    $$V = lwh$$
    $$36 \text{ in}^3 = 8 \text{ in.} \times 3 \text{ in.} \times h$$
    $$36 \text{ in}^3 = 24 \text{ in}^2 \times h$$
    $$\left(\frac{1}{24 \text{ in}^2}\right) 36 \text{ in}^3 = \left(\frac{1}{24 \text{ in}^2}\right) \times 24 \text{ in}^2 \times h$$
    $$1\frac{1}{2} \text{ in.} = h$$

    > I can use $V = lwh$ because the bases of the prism are rectangles. Using this formula, I can use my knowledge of solving equations to determine the height.

    The height of the prism is $1\frac{1}{2}$ in.

Lesson 23:   The Volume of a Right Prism

1. Calculate the volume of each solid using the formula $V = Bh$ (all angles are 90 degrees).

a.
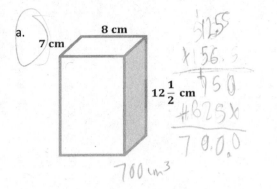

8 cm
7 cm
$12\frac{1}{2}$ cm

5125
X156.5
750
#625X
79.0.0
700 cm³

b.
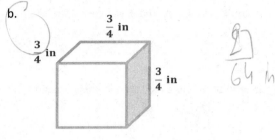

$\frac{3}{4}$ in
$\frac{3}{4}$ in
$\frac{3}{4}$ in

$\frac{9}{64}$ in³

c.
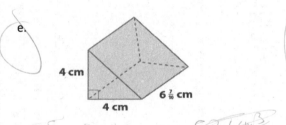

$4\frac{1}{2}$ in
4 in
$\frac{1}{2}$ in
$1\frac{1}{2}$ in
$1\frac{1}{2}$ in

d.

6 yd
$\frac{2}{3}$ yd
1 yd $1\frac{1}{3}$ yd 3 yd 4 yd

e.
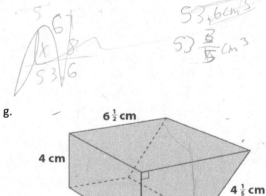

4 cm
4 cm
$6\frac{2}{10}$ cm

$5 \times 6 \div 8$
$53\frac{6}{}$

53.6 cm³
53$\frac{3}{5}$ cm³

f.
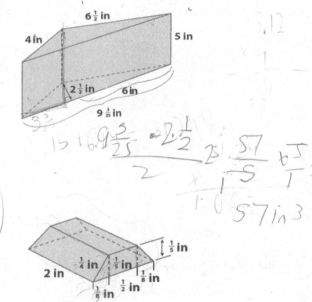

$6\frac{1}{2}$ in
4 in
5 in
$2\frac{1}{2}$ in
6 in
$9\frac{3}{25}$ in

$15 \quad 6.9\frac{3}{25} \cdot 2\frac{1}{2}$
$\frac{B \cdot \frac{57}{5}}{2} \quad \frac{6.5}{1} = 57$
1.06 57 in³

g.

$6\frac{1}{2}$ cm
4 cm
9 cm
$4\frac{1}{5}$ cm
$5\frac{1}{4}$ cm

h.

$\frac{1}{5}$ in
$\frac{1}{4}$ in $\frac{1}{5}$ in
2 in
$\frac{1}{8}$ in
$\frac{1}{8}$ in $\frac{1}{2}$ in

$\frac{1}{8} + \frac{11}{5} \cdot \frac{1}{2} \frac{3}{8} = \frac{A}{8}$
$\frac{3}{2} + \frac{B}{5}$ in $\frac{10}{8} = \frac{3}{10} = 0.3.3$

2. Let l represent the length, w the width, and h the height of a right rectangular prism. Find the volume of the prism when

   a. $l = 3$ cm, $w = 2\frac{1}{2}$ cm, and $h = 7$ cm.

   b. $l = \frac{1}{4}$ cm, $w = 4$ cm, and $h = 1\frac{1}{2}$ cm.

3. Find the length of the edge indicated in each diagram.

   a.

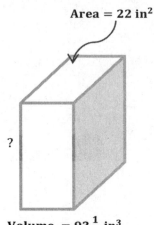

Area = 22 in²

?

Volume $= 93\frac{1}{2}$ in³

   What are possible dimensions of the base?

   b.

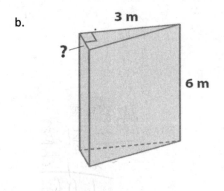

3 m

?

6 m

Volume = 4½ m³

4. The volume of a cube is $3\frac{3}{8}$ in³. Find the length of each edge of the cube.

5. Given a right rectangular prism with a volume of $7\frac{1}{2}$ ft³, a length of 5 ft, and a width of 2 ft, find the height of the prism.

© 2019 Great Minds®. eureka-math.org

EUREKA
MATH

Exploratory Challenge: Measuring a Container's Capacity

A box in the shape of a right rectangular prism has a length of 12 in, a width of 6 in, and a height of 8 in The base and the walls of the container are $\frac{1}{4}$ in. thick, and its top is open. What is the capacity of the right rectangular prism? (Hint: The capacity is equal to the volume of water needed to fill the prism to the top.)

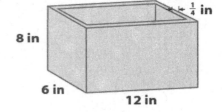

Example 1: Measuring Liquid in a Container in Three Dimensions

A glass container is in the form of a right rectangular prism. The container is 10 cm long, 8 cm wide, and 30 cm high. The top of the container is open, and the base and walls of the container are 3 mm (or 0.3 cm) thick. The water in the container is 6 cm from the top of the container. What is the volume of the water in the container?

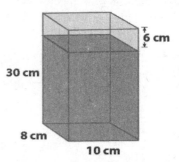

Example 2

7.2 L of water are poured into a container in the shape of a right rectangular prism. The inside of the container is 50 cm long, 20 cm wide, and 25 cm tall. How far from the top of the container is the surface of the water? ($1 \text{ L} = 1000 \text{ cm}^3$)

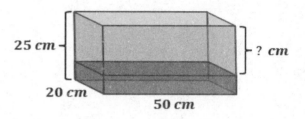

Example 3

A fuel tank is the shape of a right rectangular prism and has 27 L of fuel in it. It is determined that the tank is $\frac{3}{4}$ full. The inside dimensions of the base of the tank are 90 cm by 50 cm. What is the height of the fuel in the tank? How deep is the tank? ($1 \text{ L} = 1,000 \text{ cm}^3$)

Name _____   Date _____

Lawrence poured 27.328 L of water into a right rectangular prism-shaped tank. The base of the tank is 40 cm by 28 cm. When he finished pouring the water, the tank was $\frac{2}{3}$ full. (1 L = 1,000 cm^3)

    a.   How deep is the water in the tank?

    b.   How deep is the tank?

    c.   How many liters of water can the tank hold in total?

> If I did not know the inside dimensions, I would not be able to calculate the correct volume because the thickness of the walls has an impact on the volume when using dimensions on the outside of the tank.

1. Whitney bought an aquarium that is a right rectangular prism. The inside dimensions of the aquarium are 75 cm long, by 50 cm, by 70 cm deep. She plans to put water in the aquarium before purchasing any pet fish. How many liters of water does she need to put in the aquarium so that the water level is 8 cm from the top?

$V = lwh$

$V = 75 \text{ cm} \times 50 \text{ cm} \times 62 \text{ cm}$

$V = 232,500 \text{ cm}^3$

> The height of the water is only 62 cm because the water level is 8 cm below the top of the aquarium.

$232,500 \text{ cm}^3 = 232.5 \text{ L}$

> I know that there are 1,000 cubic centimeters in 1 liter.

The volume of the water needed is 232. 5 L.

2. The insides of two different water tanks are shown below. Which tank has the smaller capacity? Justify your answer.

$V_1 = Bh$

$V_1 = (9 \text{ ft.} \times 2.8 \text{ ft.}) \times 4 \text{ ft.}$

$V_1 = 25.2 \text{ ft}^2 \times 4 \text{ ft.}$

$V_1 = 100.8 \text{ ft}^3$

> I need to determine the volume of each tank in order to determine which one has a smaller capacity.

4 ft.   Tank 1   Tank 2   12 ft.

9 ft.   2.8 ft.   2.8 ft.   3 ft.

$V_2 = (2.8 \text{ ft.} \times 3 \text{ ft.}) \times 12 \text{ ft.}$

$V_2 = 8.4 \text{ ft}^2 \times 12 \text{ ft.}$

$V_2 = 100.8 \text{ ft}^3$

Each prism has a volume of 100.8 ft^3, which means that both tanks have the same capacity so neither one has a smaller capacity.

3. The inside base of a right rectangular prism-shaped tank is 42 cm by 59 cm. What is the minimum height inside the tank if the volume of the liquid in the tank is 74.34 L?

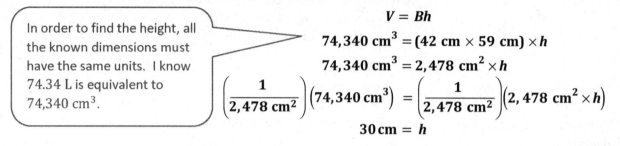

> In order to find the height, all the known dimensions must have the same units. I know 74.34 L is equivalent to 74,340 cm³.

$$V = Bh$$
$$74,340 \text{ cm}^3 = (42 \text{ cm} \times 59 \text{ cm}) \times h$$
$$74,340 \text{ cm}^3 = 2,478 \text{ cm}^2 \times h$$
$$\left(\frac{1}{2,478 \text{ cm}^2}\right)(74,340 \text{ cm}^3) = \left(\frac{1}{2,478 \text{ cm}^2}\right)(2,478 \text{ cm}^2 \times h)$$
$$30 \text{ cm} = h$$

The minimum height of the tank is 30 cm.

4. The inside of a right rectangular prism-shaped tank has a base that is 18 cm by 30 cm and a height of 42cm. The tank is filled to its capacity with water, and then 4.32 L of water was removed. How far did the water level drop?

$$V = Bh$$
$$V = (18 \text{ cm} \times 30 \text{ cm}) \times 42 \text{ cm}$$
$$V = 540 \text{ cm}^2 \times 42 \text{ cm}$$
$$V = 22,680 \text{ cm}^3$$

> I need to determine the maximum capacity of the tank before I can worry about letting water out.

The capacity of the tank is 22,680 cm³ or 22.68 L.

$$22.68 \text{ L} - 4.32 \text{ L} = 18.36 \text{ L}$$

> When 4.32 L are removed from the tank, 18.36 L, or 18,360 cm³, are left.

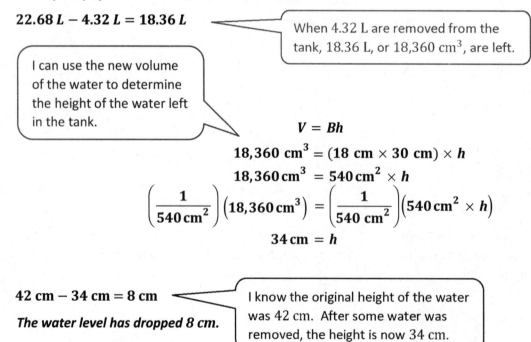

> I can use the new volume of the water to determine the height of the water left in the tank.

$$V = Bh$$
$$18,360 \text{ cm}^3 = (18 \text{ cm} \times 30 \text{ cm}) \times h$$
$$18,360 \text{ cm}^3 = 540 \text{ cm}^2 \times h$$
$$\left(\frac{1}{540 \text{ cm}^2}\right)(18,360 \text{ cm}^3) = \left(\frac{1}{540 \text{ cm}^2}\right)(540 \text{ cm}^2 \times h)$$
$$34 \text{ cm} = h$$

$$42 \text{ cm} - 34 \text{ cm} = 8 \text{ cm}$$

> I know the original height of the water was 42 cm. After some water was removed, the height is now 34 cm.

The water level has dropped 8 cm.

Lesson 24:   The Volume of a Right Prism

EUREKA MATH

© 2019 Great Minds®. eureka-math.org

5. A tank in the shape of a right rectangular prism has inside dimensions of $35\frac{1}{2}$ cm long and $42\frac{1}{5}$ cm wide. The tank is $\frac{3}{4}$ full of water. It contains 89.886 L of water. Find the depth of the container.

$$V = lwh$$

$$89{,}886 \text{ cm}^3 = \left(35\frac{1}{2} \text{ cm} \times 42\frac{1}{5} \text{ cm} \right) \times h$$

$$89{,}886 \text{ cm}^3 = 1{,}498.1 \text{ cm}^2 \times h$$

$$\left(\frac{1}{1{,}498.1 \text{ cm}^2} \right) \left(89{,}886 \text{ cm}^3 \right) = \left(\frac{1}{1{,}498.1 \text{ cm}^2} \right) \left(1{,}498.1 \text{ cm}^2 \times h \right)$$

$$60 \text{ cm} = h$$

> If I determine the height of the water, I can use this information to determine the depth of the container.

Let d represent the depth of the container in centimeters.

$$60 \text{ cm} = \frac{3}{4} \cdot d$$

$$\left(\frac{4}{3} \right)(60 \text{ cm}) = \left(\frac{4}{3} \right)\left(\frac{3}{4} \cdot d \right)$$

$$80 \text{ cm} = d$$

> I know the water only fills $\frac{3}{4}$ of the tank, so the depth of the container should be larger than the height of the water.

The depth of the container is 80 cm.

1. Mark wants to put some fish and decorative rocks in his new glass fish tank. He measured the outside dimensions of the right rectangular prism and recorded a length of 55 cm, width of 42 cm, and height of 38 cm. He calculates that the tank will hold 87.78 L of water. Why is Mark's calculation of volume incorrect? What is the correct volume? Mark also failed to take into account the fish and decorative rocks he plans to add. How will this affect the volume of water in the tank? Explain.

2. Leondra bought an aquarium that is a right rectangular prism. The inside dimensions of the aquarium are 90 cm long, by 48 cm wide, by 60 cm deep. She plans to put water in the aquarium before purchasing any pet fish. How many liters of water does she need to put in the aquarium so that the water level is 5 cm below the top?

3. The inside space of two different water tanks are shown below. Which tank has a greater capacity? Justify your answer.

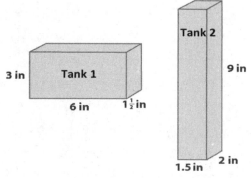

4. The inside of a tank is in the shape of a right rectangular prism. The base of that prism is 85 cm by 64 cm. What is the minimum height inside the tank if the volume of the liquid in the tank is 92 L ?

5. An oil tank is the shape of a right rectangular prism. The inside of the tank is 36.5 cm long, 52 cm wide, and 29 cm high. If 45 liters of oil have been removed from the tank since it was full, what is the current depth of oil left in the tank?

6. The inside of a right rectangular prism-shaped tank has a base that is 14 cm by 24 cm and a height of 60 cm. The tank is filled to its capacity with water, and then 10.92 L of water is removed. How far did the water level drop?

7. A right rectangular prism-shaped container has inside dimensions of $7\frac{1}{2}$ cm long and $4\frac{3}{5}$ cm wide. The tank is $\frac{3}{5}$ full of vegetable oil. It contains 0.414 L of oil. Find the height of the container.

8. A right rectangular prism with length of 10 in, width of 16 in, and height of 12 in is $\frac{2}{3}$ filled with water. If the water is emptied into another right rectangular prism with a length of 12 in., a width of 12 in., and height of 9 in., will the second container hold all of the water? Explain why or why not. Determine how far (above or below) the water level would be from the top of the container.

Opening Exercise

What is the surface area and volume of the right rectangular prism?

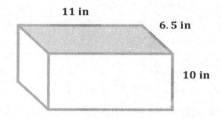

11 in

6.5 in

10 in

Example 1: Volume of a Fish Tank

Jay has a small fish tank. It is the same shape and size as the right rectangular prism shown in the Opening Exercise.

a. The box it came in says that it is a 3-gallon tank. Is this claim true? Explain your reasoning. Recall that $1 \text{ gal} = 231 \text{ in}^3$.

b. The pet store recommends filling the tank to within 1.5 in. of the top. How many gallons of water will the tank hold if it is filled to the recommended level?

c. Jay wants to cover the back, left, and right sides of the tank with a background picture. How many square inches will be covered by the picture?

d. Water in the tank evaporates each day, causing the water level to drop. How many gallons of water have evaporated by the time the water in the tank is four inches deep? Assume the tank was filled to within 1.5 in. of the top to start.

Exercise 1: Fish Tank Designs

Two fish tanks are shown below, one in the shape of a right rectangular prism (R) and one in the shape of a right trapezoidal prism (T).

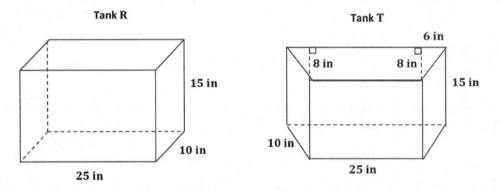

a. Which tank holds the most water? Let $Vol(R)$ represent the volume of the right rectangular prism and $Vol(T)$ represent the volume of the right trapezoidal prism. Use your answer to fill in the blanks with $Vol(R)$ and $Vol(T)$.

_____ < _____

b.   Which tank has the most surface area? Let $SA(R)$ represent the surface area of the right rectangular prism and $SA(T)$ represent the surface area of the right trapezoidal prism. Use your answer to fill in the blanks with $SA(R)$ and $SA(T)$.

　　　　_____ < _____

c.   Water evaporates from each aquarium. After the water level has dropped $\frac{1}{2}$ inch in each aquarium, how many cubic inches of water are required to fill up each aquarium? Show work to support your answers.

Exercise 2: Design Your Own Fish Tank

Design at least three fish tanks that will hold approximately 10 gallons of water. All of the tanks should be shaped like right prisms. Make at least one tank have a base that is not a rectangle. For each tank, make a sketch, and calculate the volume in gallons to the nearest hundredth.

Challenge: Each tank is to be constructed from glass that is $\frac{1}{4}$ in. thick. Select one tank that you designed, and determine the difference between the volume of the total tank (including the glass) and the volume inside the tank. Do not include a glass top on your tank.

Name _____   Date _____

Melody is planning a raised bed for her vegetable garden.

15 in

30 in

48 in

a.   How many square feet of wood does she need to create the bed?

b.   She needs to add soil. Each bag contains 1.5 cubic feet. How many bags will she need to fill the vegetable garden?

© 2019 Great Minds®. eureka-math.org

1. The dimensions of two right rectangular fish tanks are listed below. Find the volume in cubic centimeters, the capacity in liters, and the surface area in square centimeters for each tank. What do you observe about the change in volume compared with the change in surface area between the two tanks?

Tank Size	Length (cm)	Width (cm)	Height (cm)
Small	20	14	11
Large	32	20	21

$$V_S = 20 \text{ cm} \times 14 \text{ cm} \times 11 \text{ cm} = 3{,}080 \text{ cm}^3$$

> Once I know the volume, I can calculate the capacity by converting the cubic centimeters into liters.

$$V_l = 32 \text{ cm} \times 20 \text{ cm} \times 21 \text{ cm} = 13{,}440 \text{ cm}^3$$

$$SA_s = 2(20\,\text{cm} \times 14\,\text{cm}) + 2(14\,\text{cm} \times 11\,\text{cm}) + 2(20\,\text{cm} \times 11\,\text{cm})$$
$$SA_s = 2(280\,\text{cm}^2) + 2(154\,\text{cm}^2) + 2(220\,\text{cm}^2)$$
$$SA_s = 560\,\text{cm}^2 + 308\,\text{cm}^2 + 440\,\text{cm}^2$$
$$SA_s = 1{,}308\,\text{cm}^2$$

> I remember how to calculate the surface area from Lessons 21 and 22.

$$SA_l = 2(32\,\text{cm} \times 20\,\text{cm}) + 2(20\,\text{cm} \times 21\,\text{cm}) + 2(32\,\text{cm} \times 21\,\text{cm})$$
$$SA_l = 2(640\,\text{cm}^2) + 2(420\,\text{cm}^2) + 2(672\,\text{cm}^2)$$
$$SA_l = 1{,}280\,\text{cm}^2 + 840\,\text{cm}^2 + 1{,}344\,\text{cm}^2$$
$$SA_l = 3{,}464\,\text{cm}^2$$

Tank Size	Volume (cm^3)	Capacity (L)	Surface Area (cm^2)
Small	3,080	3.08	1,308
Large	13,440	13.44	3,464

The volume of the large tank is about four times the volume of the small tank. However, the surface area of the large tank is only about two and half times the surface area of the small tank.

2. A rectangular container 30 cm long by 20 cm wide contains 12 L of water.

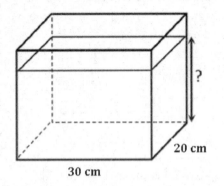

a. Find the height of the water level in the container.

$$12\,L = 12,000\text{ cm}^3$$

> I have to convert the capacity of the water to cm^3 before finding the height of the water.

$$12,000\,\text{cm}^3 = 30\,\text{cm} \times 20\,\text{cm} \times h$$
$$12,000\,\text{cm}^3 = 600\,\text{cm}^2 \times h$$
$$\left(\frac{1}{600\,\text{cm}^2}\right)(12,000\,\text{cm}^3) = \left(\frac{1}{600\,\text{cm}^2}\right)\left(600\,\text{cm}^2 \times h\right)$$
$$20\,\text{cm} = h$$

***The height of the water is* 20 cm.**

b. If the height of the container is 24 cm, how many more liters of water would it take to completely fill the container?

$$V = 30\text{ cm} \times 20\,\text{cm} \times 24\,\text{cm} = 14,400\,\text{cm}^3$$

$$14.4\,L - 12\,L = 2.4\,L$$

> The total capacity of the container is 14.4 L.

> I need to find the difference between the total capacity and the capacity of the water to determine the amount of water needed to fill the container.

To completely fill the tank,* 2.4 *L of water would have to be added to the tank.

     Lesson 25:   Volume and Surface Area

c. What percentage of the tank is filled when it contains 12 L of water?

$$\frac{12 \ L}{14.4 \ L} = 0.8\bar{3} = 83\frac{1}{3}\%$$

> I divide the part by the whole in order to find the percent.

3. Two tanks are shown below. Both are filled to capacity, but the owner decides to drain them. Tank 1 is draining at a rate of 6 liters per minute. Tank 2 is draining at a rate of 7 liters per minute. Which tank empties first?

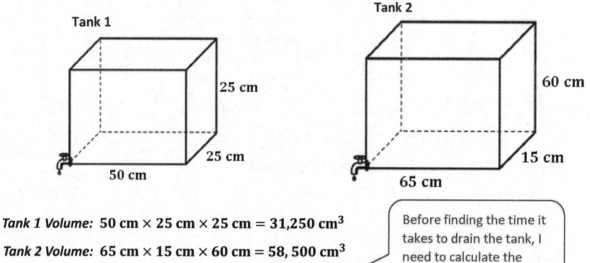

Tank 1 Volume: 50 cm × 25 cm × 25 cm = 31,250 cm³

Tank 2 Volume: 65 cm × 15 cm × 60 cm = 58,500 cm³

> Before finding the time it takes to drain the tank, I need to calculate the capacity of each tank.

Tank 1 Capacity: 31.25 L

Tank 2 Capacity: 58.5 L

Time to drain Tank 1: $\dfrac{31.25 L}{6\frac{L}{min}} \approx 5.2$ min.  **Time to drain Tank 2:** $\dfrac{58.5 L}{7\frac{L}{min}} \approx 8.4$ min.

> I divide the capacity of each tank by the rate at which the tank drains to determine how long it takes to drain each tank.

Tank 1 will be empty first because it will drain in about 5.2 minutes, and Tank 2 drains in about 8.4 minutes.

4. Two tanks have equal volumes. The tops are open. The owner wants to cover one tank with a glass top. The cost of glass is $0.08 per square inch. Which tank would be less expensive to cover? How much less?

Dimensions of Tank 1: 14 in. long by 6 in. wide by 9 in. high

Dimensions of Tank 2: 12 in. long by 9 in. wide by 7 in. high

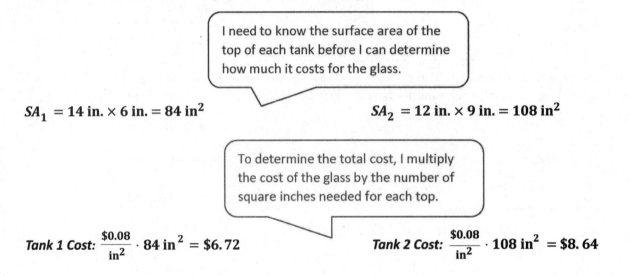

> I need to know the surface area of the top of each tank before I can determine how much it costs for the glass.

$SA_1 = 14$ in. $\times$ 6 in. $= 84$ in^2 $SA_2 = 12$ in. $\times$ 9 in. $= 108$ in^2

> To determine the total cost, I multiply the cost of the glass by the number of square inches needed for each top.

Tank 1 Cost: $\dfrac{\$0.08}{\text{in}^2} \cdot 84 \text{ in}^2 = \$6.72$ *Tank 2 Cost:* $\dfrac{\$0.08}{\text{in}^2} \cdot 108 \text{ in}^2 = \$8.64$

The second tank is $1.92 *cheaper to cover than the first tank.*

1. The dimensions of several right rectangular fish tanks are listed below. Find the volume in cubic centimeters, the capacity in liters ($1 \text{ L} = 1{,}000 \text{ cm}^3$), and the surface area in square centimeters for each tank. What do you observe about the change in volume compared with the change in surface area between the small tank and the extra-large tank?

Tank Size	Length (cm)	Width (cm)	Height (cm)
Small	24	18	15
Medium	30	21	20
Large	36	24	25
Extra-Large	40	27	30

Tank Size	Volume (cm^3)	Capacity (L)	Surface Area (cm^2)
Small			
Medium			
Large			
Extra-Large			

2. A rectangular container 15 cm long by 25 cm wide contains 2.5 L of water.

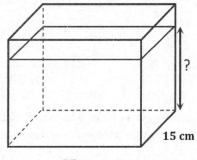

?

15 cm

25 cm

a. Find the height of the water level in the container. ($1 \text{ L} = 1{,}000 \text{ cm}^3$)

b. If the height of the container is 18 cm, how many more liters of water would it take to completely fill the container?

c. What percentage of the tank is filled when it contains 2.5 L of water?

3. A rectangular container measuring 20 cm by 14.5 cm by 10.5 cm is filled with water to its brim. If 300 cm³ are drained out of the container, what will be the height of the water level? If necessary, round to the nearest tenth.

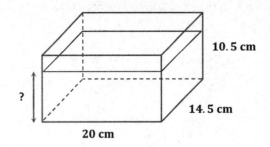

4. Two tanks are shown below. Both are filled to capacity, but the owner decides to drain them. Tank 1 is draining at a rate of 8 liters per minute. Tank 2 is draining at a rate of 10 liters per minute. Which tank empties first?

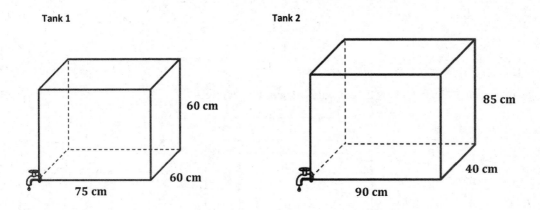

5. Two tanks are shown below. One tank is draining at a rate of 8 liters per minute into the other one, which is empty. After 10 minutes, what will be the height of the water level in the second tank? If necessary, round to the nearest minute.

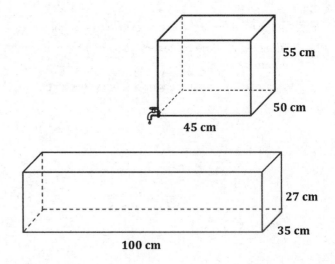

6. Two tanks with equal volumes are shown below. The tops are open. The owner wants to cover one tank with a glass top. The cost of glass is $0.05 per square inch. Which tank would be less expensive to cover? How much less?

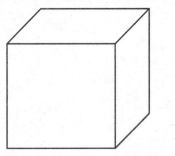

Dimensions: 12 in. long by 8 in. wide by 10 in. high   Dimensions: 15 in. long by 8 in. wide by 8 in. high

7. Each prism below is a gift box sold at the craft store.

    (a)   **(b)**

    14 cm, 8 cm, 6 cm   15 cm, 20 cm, 5 cm

    (c)   **(d)**

    8 cm, 22 cm, 5 cm, 10 cm   11 cm, 17 cm, 7 cm, 8 cm

    a. What is the volume of each prism?

    b. Jenny wants to fill each box with jelly beans. If one ounce of jelly beans is approximately 30 cm^3, estimate how many ounces of jelly beans Jenny will need to fill all four boxes? Explain your estimates.

8. Two rectangular tanks are filled at a rate of 0.5 cubic inches per minute. How long will it take each tank to be half-full?

    a. Tank 1 Dimensions: 15 in by 10 in by 12.5 in

    b. Tank 2 Dimensions: $2\frac{1}{2}$ in by $3\frac{3}{4}$ in by $4\frac{3}{8}$ in

Opening Exercise

Explain to your partner how you would calculate the area of the shaded region. Then, calculate the area.

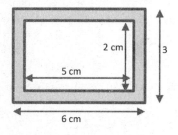

Example 1: Volume of a Shell

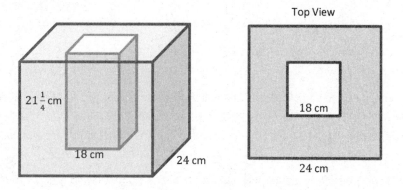

Top View

The insulated box shown is made from a large cube with a hollow inside that is a right rectangular prism with a square base. The figure on the right is what the box looks like from above.

a. Calculate the volume of the outer box.

b. Calculate the volume of the inner prism.

c. Describe in words how you would find the volume of the insulation.

d. Calculate the volume of the insulation in cubic centimeters.

e. Calculate the amount of water the box can hold in liters.

Exercise 1: Brick Planter Design

You have been asked by your school to design a brick planter that will be used by classes to plant flowers. The planter will be built in the shape of a right rectangular prism with no bottom so water and roots can access the ground beneath. The exterior dimensions are to be 12 ft. × 9 ft. × $2\frac{1}{2}$ ft. The bricks used to construct the planter are 6 in. long, $3\frac{1}{2}$ in. wide, and 2 in. high.

a. What are the interior dimensions of the planter if the thickness of the planter's walls is equal to the length of the bricks?

b. What is the volume of the bricks that form the planter?

© 2019 Great Minds®. eureka-math.org

c. If you are going to fill the planter $\frac{3}{4}$ full of soil, how much soil will you need to purchase, and what will be the height of the soil?

d. How many bricks are needed to construct the planter?

e. Each brick used in this project costs $0.82 and weighs 4.5 lb. The supply company charges a delivery fee of $15 per whole ton (2,000 lb.) over 4,000 lb. How much will your school pay for the bricks (including delivery) to construct the planter?

f.   A cubic foot of topsoil weighs between 75 and 100 lb. How much will the soil in the planter weigh?

g.   If the topsoil costs $0.88 per each cubic foot, calculate the total cost of materials that will be used to construct the planter.

Exercise 2: Design a Feeder

You did such a good job designing the planter that a local farmer has asked you to design a feeder for the animals on his farm. Your feeder must be able to contain at least 100,000 cubic centimeters, but not more than 200,000 cubic centimeters of grain when it is full. The feeder is to be built of stainless steel and must be in the shape of a right prism but not a right rectangular prism. Sketch your design below including dimensions. Calculate the volume of grain that it can hold and the amount of metal needed to construct the feeder.

The farmer needs a cost estimate. Calculate the cost of constructing the feeder if $\frac{1}{2}$ cm thick stainless steel sells for $93.25 per square meter.

Name _____   Date _____

Lawrence is designing a cooling tank that is a square prism. A pipe in the shape of a smaller 2 ft × 2 ft square prism passes through the center of the tank as shown in the diagram, through which a coolant will flow.

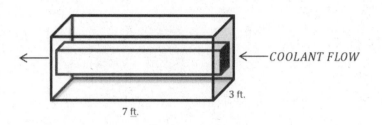

a.   What is the volume of the tank including the cooling pipe?

b.   What is the volume of coolant that fits inside the cooling pipe?

c.   What is the volume of the shell (the tank not including the cooling pipe)?

d.   Find the surface area of the cooling pipe.

1. A child's toy is constructed by cutting a right triangular prism out of a right rectangular prism. The image not drawn to scale.

Top View

a. Calculate the volume of the right rectangular prism.

$$V = 14 \text{ in.} \times 16 \text{ in.} \times 6\frac{1}{2} \text{ in.} = 1,456 \text{ in}^3$$

> I use the area formula for a triangle to calculate the area of the base of the triangular prism.

b. Calculate the volume of the triangular prism.

$$V = \frac{1}{2}\left(7 \text{ in.} \times 4 \text{ in.}\right) \times 6\frac{1}{2} \text{ in.} = 14 \text{ in}^2 \times 6\frac{1}{2} \text{ in.} = 91 \text{ in}^3$$

c. Calculate the volume of the material remaining in the rectangular prism.

$$V = 1,456 \text{ in}^3 - 91 \text{ in}^3 = 1,365 \text{ in}^3$$

> The remaining volume is the difference between the volumes of the rectangular prism and the triangular prism.

d. What is the largest number of triangular prisms that can be cut from the rectangular prism?

$$\frac{1,456 \text{ in}^3}{91 \text{ in}^3} = 16$$

> I calculate the quotient of the two volumes to determine how many triangular prisms will fit inside the rectangular prism.

e. What is the surface area of the triangular prism (assume there is no top or bottom)?

$$SA = 7 \text{ in.} \times 6\frac{1}{2} \text{ in.} + 4 \text{ in.} \times 6\frac{1}{2} \text{ in.} + 8.1 \text{ in.} \times 6\frac{1}{2} \text{ in.}$$

$$SA = 45.5 \text{ in}^2 + 26 \text{ in}^2 + 52.65 \text{ in}^2$$

$$SA = 124.15 \text{ in}^2$$

> I only have to find the area of the three faces of the triangle since there is no top or bottom.

2. A landscape designer is constructing a flower bed in the shape of a right trapezoidal prism. He needs to run three identical square prisms through the bed for drainage.

a. What is the volume of the bed without the drainage pipes?

$$V = \frac{1}{2}\left(8 \text{ in.} + 10 \text{ in.}\right) \times 5 \text{ in.} \times 12 \text{ in.} = 9 \text{ in.} \times 5 \text{ in.} \times 12 \text{ in.} = 540 \text{ in}^3$$

> The base is a trapezoid. I know the area formula for a trapezoid is $A = \frac{1}{2}(b_1 + b_2) \times h$.

b. What is the total volume of the drainage pipes?

$$V = 3(1 \text{ in}^2 \times 12 \text{ in.}) = 3(12 \text{ in}^3) = 36 \text{ in}^3$$

c. What is the volume of the soil if the planter is filled to $\frac{3}{5}$ of its total capacity with the pipes in place?

$$V = \frac{3}{5}\left(540 \text{ in}^3\right) - 36 \text{ in}^3 = 288 \text{ in}^3$$

> I know the volume of the soil is $\frac{3}{5}$ of the total volume minus the volume of the drainage pipes.

d. What is the height of the soil? If necessary, round to the nearest tenth.

$$288 \text{ in}^3 = \frac{1}{2}(8 \text{ in.} + 10 \text{ in.}) \times h \times 12 \text{ in.}$$

$$288 \text{ in}^3 = 108 \text{ in}^2 \times h$$

$$\left(\frac{1}{108 \text{ in}^2}\right)\left(288 \text{ in}^3\right) = \left(\frac{1}{108 \text{ in}^2}\right)\left(108 \text{ in}^2 \times h\right)$$

$$2.7 \text{ in.} \approx h$$

The soil has a height of about 2.7 in.

Lesson 26:   Volume and Surface Area

EUREKA MATH

1. A child's toy is constructed by cutting a right triangular prism out of a right rectangular prism.

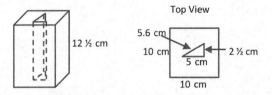

a. Calculate the volume of the rectangular prism.

b. Calculate the volume of the triangular prism.

c. Calculate the volume of the material remaining in the rectangular prism.

d. What is the largest number of triangular prisms that can be cut from the rectangular prism?

e. What is the surface area of the triangular prism (assume there is no top or bottom)?

2. A landscape designer is constructing a flower bed in the shape of a right trapezoidal prism. He needs to run three identical square prisms through the bed for drainage.

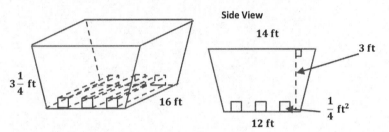

a. What is the volume of the bed without the drainage pipes?

b. What is the total volume of the three drainage pipes?

c. What is the volume of soil if the planter is filled to $\frac{3}{4}$ of its total capacity with the pipes in place?

d. What is the height of the soil? If necessary, round to the nearest tenth.

e. If the bed is made of 8 ft. × 4 ft. pieces of plywood, how many pieces of plywood will the landscape designer need to construct the bed without the drainage pipes?

f. If the plywood needed to construct the bed costs $35 per 8 ft. × 4 ft. piece, the drainage pipes cost $125 each, and the soil costs $1.25/cubic foot, how much does it cost to construct and fill the bed?

Credits

Great Minds® has made every effort to obtain permission for the reprinting of all copyrighted material. If any owner of copyrighted material is not acknowledged herein, please contact Great Minds for proper acknowledgment in all future editions and reprints of this module.